T0291693

CAMBRIDGE LIBRARY COLLECTION

Books of enduring scholarly value

Darwin

Two hundred years after his birth and 150 years after the publication of 'On the Origin of Species', Charles Darwin and his theories are still the focus of worldwide attention. This series offers not only works by Darwin, but also the writings of his mentors in Cambridge and elsewhere, and a survey of the impassioned scientific, philosophical and theological debates sparked by his 'dangerous idea'.

The Foundation of the Origin of Species

The development of Charles Darwin's views on evolution by natural selection has fascinated biologists since the 1859 publication of his landmark text On The Origin of Species. His experiences, observations and reflections during and after his pivotal journey on the Beagle during 1831–36 were of critical importance. Darwin was not, however, a man to be rushed. While his autobiography claims that the framework of his theory was laid down by 1839, its first outline sketch did not emerge until 1842. That essay was heavily edited, with many insertions and erasures. It formed the vital kernel of his more expansive but also unpolished and unpublished essay of 1844. Following careful editing by his son Francis, both essays were published in 1909, and are reproduced here. Reading these side by side, and together with the Origin, permits us to scrutinize selection and evolution truly in action.

Cambridge University Press has long been a pioneer in the reissuing of out-of-print titles from its own backlist, producing digital reprints of books that are still sought after by scholars and students but could not be reprinted economically using traditional technology. The Cambridge Library Collection extends this activity to a wider range of books which are still of importance to researchers and professionals, either for the source material they contain, or as landmarks in the history of their academic discipline.

Drawing from the world-renowned collections in the Cambridge University Library, and guided by the advice of experts in each subject area, Cambridge University Press is using state-of-the-art scanning machines in its own Printing House to capture the content of each book selected for inclusion. The files are processed to give a consistently clear, crisp image, and the books finished to the high quality standard for which the Press is recognised around the world. The latest print-on-demand technology ensures that the books will remain available indefinitely, and that orders for single or multiple copies can quickly be supplied.

The Cambridge Library Collection will bring back to life books of enduring scholarly value across a wide range of disciplines in the humanities and social sciences and in science and technology.

The Foundation of the Origin of Species

Two Essays Written in 1842 and 1844 by Charles Darwin

CHARLES DARWIN
EDITED BY FRANCIS DARWIN

CAMBRIDGE UNIVERSITY PRESS

Cambridge New York Melbourne Madrid Cape Town Singapore São Paolo Delhi

Published in the United States of America by Cambridge University Press, New York

www.cambridge.org
Information on this title: www.cambridge.org/9781108004886

This edition first published 1909
This digitally printed version 2009

ISBN 978-1-108-00488-6

THE FOUNDATIONS OF THE ORIGIN OF SPECIES

CAMBRIDGE UNIVERSITY PRESS
London: FETTER LANE, E.C.
C. F. CLAY, Manager

Edinburgh: 100, PRINCES STREET
ALSO
London: H. K. LEWIS, 136, GOWER STREET, W.C.
Berlin: A. ASHER AND CO.
Leipzig: F. A. BROCKHAUS
New York: G. P. PUTNAM'S SONS
Bombay and Calcutta: MACMILLAN AND Co., Ltd.

Charles Darwin

from a photograph by Maull & Fox. circ 1854

THE FOUNDATIONS OF THE ORIGIN OF SPECIES

TWO ESSAYS WRITTEN IN 1842 AND 1844

by

CHARLES DARWIN

Edited by his son

FRANCIS DARWIN
Honorary Fellow of Christ's College

Cambridge:
at the University Press
1909

Astronomers might formerly have said that God ordered each planet to move in its particular destiny. In same manner God orders each animal created with certain form in certain country. But how much more simple and sublime power,—let attraction act according to certain law, such are inevitable consequences,—let animal⟨s⟩ be created, then by the fixed laws of generation, such will be their successors.

From DARWIN'S *Note Book*, 1837, p. 101.

TO THE MASTER AND FELLOWS OF CHRIST'S COLLEGE, THIS BOOK IS DEDICATED BY THE EDITOR IN TOKEN OF RESPECT AND GRATITUDE

CONTENTS

ESSAY OF 1842

ESSAY OF 1844

PART I

CHAPTER I

ON THE VARIATION OF ORGANIC BEINGS UNDER DOMESTICATION; AND ON THE PRINCIPLES OF SELECTION.

CHAPTER II

ON THE VARIATION OF ORGANIC BEINGS IN A WILD STATE; ON THE NATURAL MEANS OF SELECTION; AND ON THE COMPARISON OF DOMESTIC RACES AND TRUE SPECIES.

CHAPTER III

ON THE VARIATION OF INSTINCTS AND OTHER MENTAL ATTRIBUTES UNDER DOMESTICATION AND IN A STATE OF NATURE; ON THE DIFFICULTIES IN THIS SUBJECT; AND ON ANALOGOUS DIFFICULTIES WITH RESPECT TO CORPOREAL STRUCTURES.

PART II

ON THE EVIDENCE FAVOURABLE AND OPPOSED TO THE VIEW THAT SPECIES ARE NATURALLY FORMED RACES, DESCENDED FROM COMMON STOCKS.

CHAPTER IV

CHAPTER V

GRADUAL APPEARANCE AND DISAPPEARANCE OF SPECIES.

CHAPTER VI

ON THE GEOGRAPHICAL DISTRIBUTION OF ORGANIC BEINGS IN PAST AND PRESENT TIMES.

SECTION FIRST

SECOND SECTION

SECTION THIRD

CHAPTER VII

ON THE NATURE OF THE AFFINITIES AND CLASSIFICATION OF ORGANIC BEINGS.

CHAPTER VIII

UNITY OF TYPE IN THE GREAT CLASSES; AND MORPHOLOGICAL STRUCTURES.

CHAPTER IX

ABORTIVE OR RUDIMENTARY ORGANS.

CHAPTER X

RECAPITULATION AND CONCLUSION.

INTRODUCTION

We know from the contents of Charles Darwin's Note Book of 1837 that he was at that time a convinced Evolutionist[1]. Nor can there be any doubt that, when he started on board the *Beagle*, such opinions as he had were on the side of immutability. When therefore did the current of his thoughts begin to set in the direction of Evolution?

We have first to consider the factors that made for such a change. On his departure in 1831, Henslow gave him vol. I. of Lyell's *Principles*, then just published, with the warning that he was not to believe what he read[2]. But believe he did, and it is certain (as Huxley has forcibly pointed out[3]) that the doctrine of uniformitarianism when applied to Biology leads of necessity to Evolution. If the extermination of a species is no more catastrophic than the natural death of an individual, why should the birth of a species be any more miraculous than the birth of an individual? It is quite clear that this thought was vividly present to Darwin when he was writing out his early thoughts in the 1837 Note Book[4]:—

"Propagation explains why modern animals same type as extinct, which is law almost proved.

[1] See the extracts in *Life and Letters of Charles Darwin*, ii. p. 5.

[2] The second volume,—especially important in regard to Evolution,—reached him in the autumn of 1832, as Prof. Judd has pointed out in his most interesting paper in *Darwin and Modern Science*. Cambridge, 1909.

[3] Obituary Notice of C. Darwin, *Proc. R. Soc.* vol. 44. Reprinted in Huxley's *Collected Essays*. See also *Life and Letters of C. Darwin*, ii. p. 179.

[4] See the extracts in the *Life and Letters*, ii. p. 5.

They die, without they change, like golden pippins; it is a *generation of species* like generation *of individuals.*"

"If *species* generate other *species* their race is not utterly cut off."

These quotations show that he was struggling to see in the origin of species a process just as scientifically comprehensible as the birth of individuals. They show, I think, that he recognised the two things not merely as similar but as identical.

It is impossible to know how soon the ferment of uniformitarianism began to work, but it is fair to suspect that in 1832 he had already begun to see that mutability was the logical conclusion of Lyell's doctrine, though this was not acknowledged by Lyell himself.

There were however other factors of change. In his Autobiography[1] he wrote:—"During the voyage of the *Beagle* I had been deeply impressed by discovering in the Pampean formation great fossil animals covered with armour like that on the existing armadillos; secondly, by the manner in which closely allied animals replace one another in proceeding southward over the Continent; and thirdly, by the South American character of most of the productions of the Galapagos archipelago, and more especially by the manner in which they differ slightly on each island of the group; none of the islands appearing to be very ancient in a geological sense. It was evident that such facts as these, as well as many others, could only be explained on the supposition that species gradually become modified; and the subject haunted me."

Again we have to ask: how soon did any of these influences produce an effect on Darwin's mind? Different answers have been attempted. Huxley[2] held that these facts could not have produced their essential effect until the voyage had

[1] *Life and Letters*, i. p. 82.

[2] *Obituary Notice*, *loc. cit.*

come to an end, and the "relations of the existing with the extinct species and of the species of the different geographical areas with one another were determined with some exactness." He does not therefore allow that any appreciable advance towards evolution was made during the actual voyage of the *Beagle*.

Professor Judd[1] takes a very different view. He holds that November 1832 may be given with some confidence as the "date at which Darwin commenced that long series of observations and reasonings which eventually culminated in the preparation of the *Origin of Species*."

Though I think these words suggest a more direct and continuous march than really existed between fossil-collecting in 1832 and writing the *Origin of Species* in 1859, yet I hold that it was during the voyage that Darwin's mind began to be turned in the direction of Evolution, and I am therefore in essential agreement with Prof. Judd, although I lay more stress than he does on the latter part of the voyage.

Let us for a moment confine our attention to the passage, above quoted, from the Autobiography and to what is said in the Introduction to the *Origin*, Ed. i., viz. "When on board H.M.S. 'Beagle,' as naturalist, I was much struck with certain facts in the distribution of the inhabitants of South America, and in the geological relations of the present to the past inhabitants of that continent." These words, occurring where they do, can only mean one thing,—namely that the facts suggested an evolutionary interpretation. And this being so it must be true that his thoughts *began to flow in the direction of Descent* at this early date.

I am inclined to think that the "new light which was rising in his mind[2]" had not yet attained any

[1] *Darwin and Modern Science.*
[2] Huxley, *Obituary*, p. xi.

effective degree of steadiness or brightness. I think so because in his Pocket Book under the date 1837 he wrote, "In July opened first note-book on 'transmutation of species.' Had been greatly struck *from about month of previous March*[1] on character of South American fossils, and species on Galapagos Archipelago. These facts origin (*especially latter*), of all my views." But he did not visit the Galapagos till 1835 and I therefore find it hard to believe that his evolutionary views attained any strength or permanence until at any rate quite late in the voyage. The Galapagos facts are strongly against Huxley's view, for Darwin's attention was "thoroughly aroused[2]" by comparing the birds shot by himself and by others on board. The case must have struck him at once,—without waiting for accurate determinations,—as a microcosm of evolution.

It is also to be noted, in regard to the remains of extinct animals, that, in the above quotation from his Pocket Book, he speaks of March 1837 as the time at which he began to be "greatly struck on character of South American fossils," which suggests at least that the impression made in 1832 required reinforcement before a really powerful effect was produced.

We may therefore conclude, I think, that the evolutionary current in my father's thoughts had continued to increase in force from 1832 onwards, being especially reinforced at the Galapagos in 1835 and again in 1837 when he was overhauling the results, mental and material, of his travels. And that when the above record in the Pocket Book was made he unconsciously minimised the earlier beginnings of his theorisings, and laid more stress on the recent thoughts which were

[1] In this citation the italics are mine.
[2] *Journal of Researches*, Ed. 1860, p. 394.

naturally more vivid to him. In his letter[1] to Otto Zacharias (1877) he wrote, "On my return home in the autumn of 1836, I immediately began to prepare my Journal for publication, and then saw how many facts indicated the common descent of species." This again is evidence in favour of the view that the later growths of his theory were the essentially important parts of its development.

In the same letter to Zacharias he says, "When I was on board the *Beagle* I believed in the permanence of species, but as far as I can remember vague doubts occasionally flitted across my mind." Unless Prof. Judd and I are altogether wrong in believing that late or early in the voyage (it matters little which) a definite approach was made to the evolutionary standpoint, we must suppose that in 40 years such advance had shrunk in his recollection to the dimensions of "vague doubts." The letter to Zacharias shows, I think, some forgetting of the past where the author says, "But I did not become convinced that species were mutable until, I think, two or three years had elapsed." It is impossible to reconcile this with the contents of the evolutionary Note Book of 1837. I have no doubt that in his retrospect he felt that he had not been "convinced that species were mutable" until he had gained a clear conception of the mechanism of natural selection, *i.e.* in 1838—9.

But even on this last date there is some room, not for doubt, but for surprise. The passage in the Autobiography[2] is quite clear, namely that in October 1838 he read Malthus's *Essay on the principle of Population* and "being well prepared to appreciate the struggle for existence..., it at once struck me that under these circumstances favourable variations would tend to be preserved,

[1] F. Darwin's *Life of Charles Darwin* (in one volume), 1892, p. 166.
[2] *Life and Letters*, i. p. 83.

and unfavourable ones to be destroyed. The result of this would be the formation of new species. Here then I had at last got a theory by which to work."

It is surprising that Malthus should have been needed to give him the clue, when in the Note Book of 1837 there should occur—however obscurely expressed—the following forecast[1] of the importance of the survival of the fittest. "With respect to extinction, we can easily see that a variety of the ostrich (Petise[2]), may not be well adapted, and thus perish out; or on the other hand, like Orpheus[3], being favourable, many might be produced. This requires the principle that the permanent variations produced by confined breeding and changing circumstances are continued and produce(d) according to the adaptation of such circumstances, and therefore that death of species is a consequence (contrary to what would appear in America) of non-adaptation of circumstances."

I can hardly doubt, that with his knowledge of the interdependence of organisms and the tyranny of conditions, his experience would have crystallized out into "a theory by which to work" even without the aid of Malthus.

In my father's Autobiography[4] he writes, "In June 1842 I first allowed myself the satisfaction of writing a very brief abstract of my theory in pencil in 35 pages; and this was enlarged during the summer of 1844 into one of 230 pages[5], which I had fairly copied out and still possess." These two Essays, of 1842 and 1844, are now printed under the title *The Foundations of the Origin of Species.*

[1] *Life and Letters*, ii. p. 8. [2] Avestruz Petise; *i.e. Rhea Darwini.*
[3] A bird.
[4] *Life and Letters*, i. p. 84.
[5] It contains as a fact 231 pp. It is a strongly bound folio, interleaved with blank pages, as though for notes and additions. His own MS. from which it was copied contains 189 pp.

It will be noted that in the above passage he does not mention the MS. of 1842 as being in existence, and when I was at work on *Life and Letters* I had not seen it. It only came to light after my mother's death in 1896 when the house at Down was vacated. The MS. was hidden in a cupboard under the stairs which was not used for papers of any value, but rather as an overflow for matter which he did not wish to destroy.

The statement in the Autobiography that the MS. was written in 1842 agrees with an entry in my father's Diary:—

"1842. May 18th went to Maer. June 15th to Shrewsbury, and on 18th to Capel Curig....During my stay at Maer and Shrewsbury (five years after commencement) wrote pencil sketch of my species theory." Again in a letter to Lyell (June 18, 1858) he speaks of his "MS. sketch written out in 1842[1]." In the *Origin of Species*, Ed. i. p. 1, he speaks of beginning his speculations in 1837 and of allowing himself to draw up some "short notes" after "five years' work," *i.e.* in 1842. So far there seems no doubt as to 1842 being the date of the first sketch; but there is evidence in favour of an earlier date[2]. Thus across the Table of Contents of the bound copy of the 1844 MS. is written in my father's hand "This was sketched in 1839." Again in a letter to Mr Wallace[3] (Jan. 25, 1859) he speaks of his own contributions to the Linnean paper[4] of July 1, 1858, as "written in 1839, now just twenty years ago." This statement as it stands is undoubtedly incorrect, since the extracts are from the MS. of 1844, about the date of which no doubt exists; but even if it could be supposed to refer to the 1842 Essay, it must, I think, be rejected. I can only account for his mistake by the supposition that my father had in

[1] *Life and Letters*, ii. p. 116.
[2] *Life and Letters*, ii. p. 10.
[3] *Life and Letters*, ii. p. 146.
[4] *J. Linn. Soc. Zool.* iii. p. 45.

mind the date (1839) at which the framework of his theory was laid down. It is worth noting that in his Autobiography (p. 88) he speaks of the time "about 1839, when the theory was clearly conceived." However this may be there can be no doubt that 1842 is the correct date. Since the publication of *Life and Letters* I have gained fresh evidence on this head. A small packet containing 13 pp. of MS. came to light in 1896. On the outside is written "First Pencil Sketch of Species Theory. Written at Maer and Shrewsbury during May and June 1842." It is not however written in pencil, and it consists of a single chapter on *The Principles of Variation in Domestic Organisms.* A single unnumbered page is written in pencil, and is headed "Maer, May 1842, useless"; it also bears the words "This page was thought of as introduction." It consists of the briefest sketch of the geological evidence for evolution, together with words intended as headings for discussion,—such as "Affinity,—unity of type,—fœtal state,—abortive organs."

The back of this "useless" page is of some interest, although it does not bear on the question of date,—the matter immediately before us.

It seems to be an outline of the Essay or sketch of 1842, consisting of the titles of the three chapters of which it was to have consisted.

"I. The Principles of Var. in domestic organisms.

"II. The possible and probable application of these same principles to wild animals and consequently the possible and probable production of wild races, analogous to the domestic ones of plants and animals.

"III. The reasons for and against believing that such races have really been produced, forming what are called species."

It will be seen that Chapter III as originally

designed corresponds to Part II (p. 22) of the Essay of 1842, which is (p. 7) defined by the author as discussing "whether the characters and relations of animated things are such as favour the idea of wild species being races descended from a common stock." Again at p. 23 the author asks "What then is the evidence in favour of it ⟨the theory of descent⟩ and what the evidence against it." The generalised section of his Essay having been originally Chapter III[1] accounts for the curious error which occurs in pp. 18 and 22 where the second Part of the Essay is called Part III.

The division of the Essay into two parts is maintained in the enlarged Essay of 1844, in which he writes: "The Second Part of this work is devoted to the general consideration of how far the general economy of nature justifies or opposes the belief that related species and genera are descended from common stocks." The *Origin of Species* however is not so divided.

We may now return to the question of the date of the Essay. I have found additional evidence in favour of 1842 in a sentence written on the back of the Table of Contents of the 1844 MS.—not the copied version but the original in my father's writing: "This was written and enlarged from a sketch in 37 pages[2] in Pencil (the latter written in summer of 1842 at Maer and Shrewsbury) in beginning of 1844, and finished it ⟨sic⟩ in July; and finally corrected the copy by Mr Fletcher in the last week in September." On the whole it is impossible to doubt that 1842 is the date of the earlier of the two Essays.

[1] It is evident that *Parts* and *Chapters* were to some extent interchangeable in the author's mind, for p. 1 (of the MS. we have been discussing) is headed in ink Chapter I, and afterwards altered in pencil to Part I.

[2] On p. 23 of the MS. of the *Foundations* is a reference to the "back of p. 21 bis": this suggests that additional pages had been interpolated in the MS. and that it may once have had 37 in place of 35 pp.

The sketch of 1842 is written on bad paper with a soft pencil, and is in many parts extremely difficult to read, many of the words ending in mere scrawls and being illegible without context. It is evidently written rapidly, and is in his most elliptical style, the articles being frequently omitted, and the sentences being loosely composed and often illogical in structure. There is much erasure and correction, apparently made at the moment of writing, and the MS. does not give the impression of having been re-read with any care. The whole is more like hasty memoranda of what was clear to himself, than material for the convincing of others.

Many of the pages are covered with writing on the back, an instance of his parsimony in the matter of paper[1]. This matter consists partly of passages marked for insertion in the text, and these can generally (though by no means always) be placed where he intended. But he also used the back of one page for a preliminary sketch to be rewritten on a clean sheet. These parts of the work have been printed as footnotes, so as to allow what was written on the front of the pages to form a continuous text. A certain amount of repetition is unavoidable, but much of what is written on the backs of the pages is of too much interest to be omitted. Some of the matter here given in footnotes may, moreover, have been intended as the final text and not as the preliminary sketch.

When a word cannot be deciphered, it is replaced by:—⟨illegible⟩, the angular brackets being, as already explained, a symbol for an insertion by the editor. More commonly, however, the context makes the interpretation of a word reasonably sure although the word is not strictly legible. Such words are followed by an inserted mark of interrogation ⟨?⟩.

[1] *Life and Letters*, i. p. 153.

Lastly, words inserted by the editor, of which the appropriateness is doubtful, are printed thus ⟨variation?⟩.

Two kinds of erasure occur in the MS. of 1842. One by vertical lines which seem to have been made when the 35 pp. MS. was being expanded into that of 1844, and merely imply that such a page is done with: and secondly the ordinary erasures by horizontal lines. I have not been quite consistent in regard to these: I began with the intention of printing (in square brackets) all such erasures. But I ultimately found that the confusion introduced into the already obscure sentences was greater than any possible gain; and many such erasures are altogether omitted. In the same way I have occasionally omitted hopelessly obscure and incomprehensible fragments, which if printed would only have burthened the text with a string of ⟨illegible⟩s and queried words. Nor have I printed the whole of what is written on the backs of the pages, where it seemed to me that nothing but unnecessary repetition would have been the result.

In the matter of punctuation I have given myself a free hand. I may no doubt have misinterpreted the author's meaning in so doing, but without such punctuation, the number of repellantly crabbed sentences would have been even greater than at present. In dealing with the Essay of 1844, I have corrected some obvious slips without indicating such alterations, because the MS. being legible, there is no danger of changing the author's meaning.

The sections into which the Essay of 1842 is divided are in the original merely indicated by a gap in the MS. or by a line drawn across the page. No titles are given except in the case of § VIII.; and § II. is the only section which has a number in the original. I might equally well have made sections of what are now subsections, *e.g. Natural*

Selection p. 7, or *Extermination* p. 28. But since the present sketch is the germ of the Essay of 1844, it seemed best to preserve the identity between the two works, by using such of the author's divisions as correspond to the chapters of the enlarged version of 1844. The geological discussion with which Part II begins corresponds to two chapters (IV and V) of the 1844 Essay. I have therefore described it as §§ IV. and V., although I cannot make sure of its having originally consisted of two sections. With this exception the ten sections of the Essay of 1842 correspond to the ten chapters of that of 1844.

The *Origin of Species* differs from the sketch of 1842 in not being divided into two parts. But the two volumes resemble each other in general structure. Both begin with a statement of what may be called the mechanism of evolution,—variation and selection: in both the argument proceeds from the study of domestic organisms to that of animals and plants in a state of nature. This is followed in both by a discussion of the *Difficulties on Theory* and this by a section *Instinct* which in both cases is treated as a special case of difficulty.

If I had to divide the *Origin* (first edition) into two parts without any knowledge of earlier MS., I should, I think, make Part II begin with Ch. VI, *Difficulties on Theory.* A possible reason why this part of the argument is given in Part I of the Essay of 1842 may be found in the Essay of 1844, where it is clear that the chapter on instinct is placed in Part I because the author thought it of importance to show that heredity and variation occur in mental attributes. The whole question is perhaps an instance of the sort of difficulty which made the author give up the division of his argument into two Parts when he wrote the *Origin.* As matters stand §§ IV. and V. of the 1842 Essay correspond to the geological chapters, IX and X, in the *Origin*. From this point onwards the material is grouped in the same order in both works: geographical distribution; affinities and classification; unity of type and morphology; abortive or rudimentary organs; recapitulation and conclusion.

In enlarging the Essay of 1842 into that of 1844, the author retained the sections of the sketch as chapters in the completer presentment. It follows that what has been said of the relation of the earlier Essay to the *Origin* is generally true of the 1844 Essay. In the latter, however, the geological discussion is, clearly instead of obscurely, divided into two chapters, which correspond roughly with Chapters IX and X of the *Origin*. But part of the contents of Chapter X (*Origin*) occurs in Chapter VI (1844) on Geographical Distribution. The treatment of distribution is particularly full and interesting in the 1844 Essay, but the arrangement of the material, especially the introduction of § III. p. 183, leads to some repetition which is avoided in the *Origin*. It should be noted that Hybridism, which has a separate chapter (VIII) in the *Origin*, is treated in Chapter II of the Essay. Finally that Chapter XIII (*Origin*) corresponds to Chapters VII, VIII and IX of the work of 1844.

The fact that in 1842, seventeen years before the publication of the *Origin*, my father should have been able to write out so full an outline of his future work, is very remarkable. In his Autobiography[1] he writes of the 1844 Essay, "But at that time I overlooked one problem of great importance....This problem is the tendency in organic beings descended from the same stock to diverge in character as they become modified." The absence of the principle of divergence is of course also a characteristic of the

[1] *Life and Letters*, i. p. 84.

sketch of 1842. But at p. 37, the author is not far from this point of view. The passage referred to is: "If any species, *A*, in changing gets an advantage and that advantage...is inherited, *A* will be the progenitor of several genera or even families in the hard struggle of nature. *A* will go on beating out other forms, it might come that *A* would people ⟨the⟩ earth,—we may now not have one descendant on our globe of the one or several original creations[1]." But if the descendants of *A* have peopled the earth by beating out other forms, they must have diverged in occupying the innumerable diverse modes of life from which they expelled their predecessors. What I wrote[2] on this subject in 1887 is I think true: "Descent with modification implies divergence, and we become so habituated to a belief in descent, and therefore in divergence, that we do not notice the absence of proof that divergence is in itself an advantage."

The fact that there is no set discussion on the principle of divergence in the 1844 Essay, makes it clear why the joint paper read before the Linnean Society on July 1, 1858, included a letter[3] to Asa Gray, as well as an extract[4] from the Essay of 1844. It is clearly because the letter to Gray includes a discussion on divergence, and was thus, probably, the only document, including this subject, which could be appropriately made use of. It shows once more how great was the importance attached by its author to the principle of divergence.

I have spoken of the hurried and condensed manner in which the sketch of 1842 is written; the style of the later Essay (1844) is more finished.

[1] In the footnotes to the Essay of 1844 attention is called to similar passages.

[2] *Life and Letters*, ii. p. 15.

[3] The passage is given in the *Life and Letters*, ii. p. 124.

[4] The extract consists of the section on *Natural Means of Selection*, p. 87.

It has, however, the air of an uncorrected MS. rather than of a book which has gone through the ordeal of proof sheets. It has not all the force and conciseness of the *Origin*, but it has a certain freshness which gives it a character of its own. It must be remembered that the *Origin* was an abstract or condensation of a much bigger book, whereas the Essay of 1844 was an expansion of the sketch of 1842. It is not therefore surprising that in the *Origin* there is occasionally evident a chafing against the author's self-imposed limitation. Whereas in the 1844 Essay there is an air of freedom, as if the author were letting himself go, rather than applying the curb. This quality of freshness and the fact that some questions were more fully discussed in 1844 than in 1859, makes the earlier work good reading even to those who are familiar with the *Origin*.

The writing of this Essay "during the summer of 1844," as stated in the Autobiography[1], and "from memory," as Darwin says elsewhere[2], was a remarkable achievement, and possibly renders more conceivable the still greater feat of the writing of the *Origin* between July 1858 and September 1859.

It is an interesting subject for speculation: what influence on the world the Essay of 1844 would have exercised, had it been published in place of the *Origin*. The author evidently thought of its publication in its present state as an undesirable expedient, as appears clearly from the following extracts from the *Life and Letters*, vol. ii. pp. 16—18:

[1] *Life and Letters*, i. p. 84. [2] *Life and Letters*, ii. p. 18.

C. Darwin to Mrs Darwin.

Down, *July* 5, 1844.

"...I have just finished my sketch of my species theory. If, as I believe, my theory in time be accepted even by one competent judge, it will be a considerable step in science.

"I therefore write this in case of my sudden death, as my most solemn and last request, which I am sure you will consider the same as if legally entered in my will, that you will devote £400 to its publication, and further will yourself, or through Hensleigh[1], take trouble in promoting it. I wish that my sketch be given to some competent person, with this sum to induce him to take trouble in its improvement and enlargement. I give to him all my books on Natural History, which are either scored or have references at the end to the pages, begging him carefully to look over and consider such passages as actually bearing, or by possibility bearing, on this subject. I wish you to make a list of all such books as some temptation to an editor. I also request that you will hand over ⟨to⟩ him all those scraps roughly divided into eight or ten brown paper portfolios. The scraps, with copied quotations from various works, are those which may aid my editor. I also request that you, or some amanuensis, will aid in deciphering any of the scraps which the editor may think possibly of use. I leave to the editor's judgment whether to interpolate these facts in the text, or as notes, or under appendices. As the looking over the references and scraps will be a long labour, and as the *correcting* and enlarging and altering my sketch will also take considerable time, I leave this sum of £400 as some remuneration, and any profits from

[1] Mrs Darwin's brother.

the work. I consider that for this the editor is bound to get the sketch published either at a publisher's or his own risk. Many of the scraps in the portfolios contain mere rude suggestions and early views, now useless, and many of the facts will probably turn out as having no bearing on my theory.

"With respect to editors, Mr Lyell would be the best if he would undertake it; I believe he would find the work pleasant, and he would learn some facts new to him. As the editor must be a geologist as well as a naturalist, the next best editor would be Professor Forbes of London. The next best (and quite best in many respects) would be Professor Henslow. Dr Hooker would be *very* good. The next, Mr Strickland[1]. If none of these would undertake it, I would request you to consult with Mr Lyell, or some other capable man, for some editor, a geologist and naturalist. Should one other hundred pounds make the difference of procuring a good editor, I request earnestly that you will raise £500.

"My remaining collections in Natural History may be given to any one or any museum where ⟨they⟩ would be accepted...."

⟨The following note seems to have formed part of the original letter, but may have been of later date:⟩

"Lyell, especially with the aid of Hooker (and of any good zoological aid), would be best of all. Without an editor will pledge himself to give up time to it, it would be of no use paying such a sum.

"If there should be any difficulty in getting an editor who would go thoroughly into the subject,

[1] After Mr Strickland's name comes the following sentence, which has been erased, but remains legible. "Professor Owen would be very good; but I presume he would not undertake such a work."

and think of the bearing of the passages marked in the books and copied out of scraps of paper, then let my sketch be published as it is, stating that it was done several years ago[1], and from memory without consulting any works, and with no intention of publication in its present form."

The idea that the sketch of 1844 might remain, in the event of his death, as the only record of his work, seems to have been long in his mind, for in August, 1854, when he had finished with the Cirripedes, and was thinking of beginning his "species work," he added on the back of the above letter, "Hooker by far best man to edit my species volume. August 1854."

I have called attention in footnotes to many points in which the *Origin* agrees with the *Foundations*. One of the most interesting is the final sentence, practically the same in the Essays of 1842 and 1844, and almost identical with the concluding words of the *Origin*. I have elsewhere pointed out[2] that the ancestry of this eloquent passage may be traced one stage further back,—to the Note Book of 1837. I have given this sentence as an appropriate motto for the *Foundations* in its character of a study of general laws. It will be remembered that a corresponding motto from Whewell's *Bridgewater Treatise* is printed opposite the title-page of the *Origin of Species*.

Mr Huxley who, about the year 1887, read the Essay of 1844, remarked that "much more weight is attached to the influence of external conditions in producing variation and to the inheritance of acquired habits than in the *Origin*." In the *Foundations* the effect of conditions is frequently mentioned, and Darwin seems to have had constantly

[1] The words "several years ago, and" seem to have been added at a later date.

[2] *Life and Letters*, ii. p. 9.

in mind the need of referring each variation to a cause. But I gain the impression that the slighter prominence given to this view in the *Origin* was not due to change of opinion, but rather because he had gradually come to take this view for granted; so that in the scheme of that book, it was overshadowed by considerations which then seemed to him more pressing. With regard to the inheritance of acquired characters I am not inclined to agree with Huxley. It is certain that the *Foundations* contains strong recognition of the importance of germinal variation, that is of external conditions acting indirectly through the "reproductive functions." He evidently considered this as more important than the inheritance of habit or other acquired peculiarities.

Another point of interest is the weight he attached in 1842—4 to "sports" or what are now called "mutations." This is I think more prominent in the *Foundations* than in the first edition of the *Origin*, and certainly than in the fifth and sixth editions.

Among other interesting points may be mentioned the "good effects of crossing" being "possibly analogous to good effects of change in condition,"—a principle which he upheld on experimental grounds in his *Cross and Self-Fertilisation* in 1876.

In conclusion, I desire to express my thanks to Mr Wallace for a footnote he was good enough to supply: and to Professor Bateson, Sir W. Thiselton-Dyer, Dr Gadow, Professor Judd, Dr Marr, Col. Prain and Dr Stapf for information on various points. I am also indebted to Mr Rutherford, of the University Library, for his careful copy of the manuscript of 1842.

CAMBRIDGE,
June 9, 1909.

EXPLANATION OF SIGNS, &c.

[ ] Means that the words so enclosed are erased in the original MS.

⟨ ⟩ Indicates an insertion by the Editor.

Origin, Ed. vi. refers to the Popular Edition.

PART I.

§ I. ⟨On Variation under Domestication, and on the Principles of Selection.⟩

An individual organism placed under new conditions [often] sometimes varies in a small degree and in very trifling respects such as stature, fatness, sometimes colour, health, habits in animals and probably disposition. Also habits of life develope certain parts. Disuse atrophies. [Most of these slight variations tend to become hereditary.]

When the individual is multiplied for long periods by buds the variation is yet small, though greater and occasionally a single bud or individual departs widely from its type (example)[1] and continues steadily to propagate, by buds, such new kind.

When the organism is bred for several generations under new or varying conditions, the variation is greater in amount and endless in kind [especially[2] holds good when individuals have long been exposed to new conditions]. The nature of the external conditions tends to effect some definite change in all or greater part of offspring,—little food, small size—certain foods harmless &c. &c. organs affected and diseases—extent unknown. A certain degree of

[1] Evidently a memorandum that an example should be given.

[2] The importance of exposure to new conditions for several generations is insisted on in the *Origin*, Ed. i. p. 7, also p. 131. In the latter passage the author guards himself against the assumption that variations are "due to chance," and speaks of "our ignorance of the cause of each particular variation." These statements are not always remembered by his critics.

variation (Müller's twins)[1] seems inevitable effect of process of reproduction. But more important is that simple ⟨?⟩ generation, especially under new conditions [when no crossing] ⟨causes⟩ infinite variation and not direct effect of external conditions, but only in as much as it affects the reproductive functions[2]. There seems to be no part (*beau idéal* of liver)[3] of body, internal or external, or mind or habits, or instincts which does not vary in some small degree and [often] some ⟨?⟩ to a great amount.

[All such] variations [being congenital] or those very slowly acquired of all kinds [decidedly evince a tendency to become hereditary], when not so become simple variety, when it does a race. Each[4] parent transmits its peculiarities, therefore if varieties allowed freely to cross, except by the *chance* of two characterized by same peculiarity happening to marry, such varieties will be constantly demolished[5]. All bisexual animals must cross, hermaphrodite plants do cross, it seems very possible that her-

[1] Cf. *Origin*, Ed. i. p. 10, vi. p. 9, "Young of the same litter, sometimes differ considerably from each other, though both the young and the parents, as Müller has remarked, have apparently been exposed to exactly the same conditions of life."

[2] This is paralleled by the conclusion in the *Origin*, Ed. i. p. 8, that "the most frequent cause of variability may be attributed to the male and female reproductive elements having been affected prior to the act of conception."

[3] The meaning seems to be that there must be some variability in the liver otherwise anatomists would not speak of the 'beau idéal' of that organ.

[4] The position of the following passage is uncertain. "If individuals of two widely different varieties be allowed to cross, a third race will be formed—a most fertile source of the variation in domesticated animals. ⟨In the *Origin*, Ed. i. p. 20 the author says that "the possibility of making distinct races by crossing has been greatly exaggerated."⟩ If freely allowed, the characters of pure parents will be lost, number of races thus ⟨illegible⟩ but differences ⟨?⟩ besides the ⟨illegible⟩. But if varieties differing in very slight respects be allowed to cross, such small variation will be destroyed, at least to our senses,—a variation [clearly] just to be distinguished by long legs will have offspring not to be so distinguished. Free crossing great agent in producing uniformity in any breed. Introduce tendency to revert to parent form."

[5] The swamping effect of intercrossing is referred to in the *Origin*, Ed. i. p. 103, vi. p. 126.

maphrodite animals do cross,—conclusion strengthened: ill effects of breeding in and in, good effects of crossing possibly analogous to good effects of change in condition ⟨?⟩[1].

Therefore if in any country or district all animals of one species be allowed freely to cross, any small tendency in them to vary will be constantly counteracted. Secondly reversion to parent form—analogue of *vis medicatrix*[2]. But if man selects, then new races rapidly formed,—of late years systematically followed,—in most ancient times often practically followed[3]. By such selection make race-horse, dray-horse—one cow good for tallow, another for eating &c.—one plant's good lay ⟨illegible⟩ in leaves another in fruit &c. &c.: the same plant to supply his wants at different times of year. By former means animals become adapted, as a direct effect to a cause, to external conditions, as size of body to amount of food. By this latter means they may also be so adapted, but further they may be adapted to ends and pursuits, which by no possibility can affect growth, as existence of tallow-chandler cannot tend to make fat. In such selected races, if not removed to new conditions, and ⟨if⟩ preserved from all cross, after several generations become very true, like each other and not varying. But man[4] selects only ⟨?⟩ what is useful and curious—has bad judgment, is capricious,—grudges to destroy those that do not come up to his pattern,—has no

[1] A discussion on the intercrossing of hermaphrodites in relation to Knight's views occurs in the *Origin*, Ed. i. p. 96, vi. p. 119. The parallelism between crossing and changed conditions is briefly given in the *Origin*, Ed. i. p. 267, vi. p. 391, and was finally investigated in *The Effects of Cross and Self-Fertilisation in the Vegetable Kingdom*, 1876.

[2] There is an article on the *vis medicatrix* in Brougham's *Dissertations*, 1839, a copy of which is in the author's library.

[3] This is the classification of selection into methodical and unconscious given in the *Origin*, Ed. i. p. 33, vi. p. 38.

[4] This passage, and a similar discussion on the power of the Creator (p. 6), correspond to the comparison between the selective capacities of man and nature, in the *Origin*, Ed. i. p. 83, vi. p. 102.

[knowledge] power of selecting according to internal variations,—can hardly keep his conditions uniform,—[cannot] does not select those best adapted to the conditions under which ⟨the⟩ form ⟨?⟩ lives, but those most useful to him. This might all be otherwise.

§ II. ⟨On Variation in a State of Nature and on the Natural Means of Selection.⟩

Let us see how far above principles of variation apply to wild animals. Wild animals vary exceedingly little—yet they are known as individuals[1]. British Plants, in many genera number quite uncertain of varieties and species: in shells chiefly external conditions[2]. Primrose and cowslip. Wild animals from different [countries can be recognized]. Specific character gives some organs as varying. Variations analogous in kind, but less in degree with domesticated animals—chiefly external and less important parts.

Our experience would lead us to expect that any and every one of these organisms would vary if ⟨the organism were⟩ taken away ⟨?⟩ and placed under new conditions. Geology proclaims a constant round of change, bringing into play, by every possible ⟨?⟩ change of climate and the death of pre-existing inhabitants, endless variations of new conditions. These ⟨?⟩ generally very slow, doubtful though ⟨illegible⟩ how far the slowness ⟨?⟩ would produce tendency to vary. But Geolog⟨ists⟩ show change in configuration which, together with the accidents of air and water and the means of transportal which every being possesses, must occasionally bring, rather suddenly, organism to new conditions and ⟨?⟩ expose it for several generations.

[1] i.e. they are individually distinguishable.
[2] See *Origin*, Ed. i. p. 133, vi. p. 165.

Hence ⟨?⟩ we should expect every now and then a wild form to vary[1]; possibly this may be cause of some species varying more than others.

According to nature of new conditions, so we might expect all or majority of organisms born under them to vary in some definite way. Further we might expect that the mould in which they are cast would likewise vary in some small degree. But is there any means of selecting those offspring which vary in the same manner, crossing them and keeping their offspring separate and thus producing selected races: otherwise as the wild animals freely cross, so must such small heterogeneous varieties be constantly counter-balanced and lost, and a uniformity of character [kept up] preserved. The former variation as the direct and necessary effects of causes, which we can see can act on them, as size of body from amount of food, effect of certain kinds of food on certain parts of bodies &c. &c.; such new varieties may then become adapted to those external [natural] agencies which act on them. But can varieties be produced adapted to end, which cannot possibly influence their structure and which it is absurd to look ⟨at⟩ as effects of chance. Can varieties like some vars. of domesticated animals, like almost all wild species be produced adapted by exquisite means to prey on one animal or to escape from another,—or rather, as it puts out of question effects of intelligence and habits, can a plant become adapted to animals, as a plant which cannot be impregnated without agency of insect; or hooked seeds depending on animal's existence: woolly animals cannot have any direct effect on seeds of plant. This point which all theories about

[1] When the author wrote this sketch he seems not to have been so fully convinced of the general occurrence of variation in nature as he afterwards became. The above passage in the text possibly suggests that at this time he laid more stress on *sports* or *mutations* than was afterwards the case.

climate adapting woodpecker[1] to crawl ⟨?⟩ up trees, ⟨illegible⟩ miseltoe, ⟨sentence incomplete⟩. But if every part of a plant or animal was to vary ⟨illegible⟩, and if a being infinitely more sagacious than man (not an omniscient creator) during thousands and thousands of years were to select all the variations which tended towards certain ends ([or were to produce causes ⟨?⟩ which tended to the same end]), for instance, if he foresaw a canine animal would be better off, owing to the country producing more hares, if he were longer legged and keener sight,—greyhound produced[2]. If he saw that aquatic ⟨animal would need⟩ skinned toes. If for some unknown cause he found it would advantage a plant, which ⟨?⟩ like most plants is occasionally visited by bees &c.: if that plant's seed were occasionally eaten by birds and were then carried on to rotten trees, he might select trees with fruit more agreeable to such birds as perched, to ensure their being carried to trees; if he perceived those birds more often dropped the seeds, he might well have selected a bird who would ⟨illegible⟩ rotten trees or [gradually select plants which ⟨he⟩ had proved to live on less and less rotten trees]. Who, seeing how plants vary in garden, what blind foolish man has done[3] in a few years, will deny an all-seeing being in thousands of years could effect (if the Creator chose to do so), either by his own direct foresight or by intermediate means,—which will represent ⟨?⟩ the creator of this universe. Seems usual means. Be it remembered I have nothing to say about life and mind and *all*

[1] The author may possibly have taken the case of the woodpecker from Buffon, *Histoire Nat. des Oiseaux*, T. vii. p. 3, 1780, where however it is treated from a different point of view. He uses it more than once, see for instance *Origin*, Ed. i. pp. 3, 60, 184, vi. pp. 3, 76, 220. The passage in the text corresponds with a discussion on the woodpecker and the mistletoe in *Origin*, Ed. i. p. 3, vi. p. 3.

[2] This illustration occurs in the *Origin*, Ed. i. pp. 90, 91, vi. pp. 110, 111.

[3] See *Origin*, Ed. i. p. 83, vi. p. 102, where the word *Creator* is replaced by *Nature*.

forms descending from one common type[1]. I speak of the variation of the existing great divisions of the organised kingdom, how far I would go, hereafter to be seen.

Before considering whether ⟨there⟩ be any natural means of selection, and secondly (which forms the 2nd Part of this sketch) the far more important point whether the characters and relations of animated ⟨things⟩ are such as favour the idea of wild species being races ⟨?⟩ descended from a common stock, as the varieties of potato or dahlia or cattle having so descended, let us consider probable character of [selected races] wild varieties.

Natural Selection. De Candolle's war of nature,—seeing contented face of nature,—may be well at first doubted; we see it on borders of perpetual cold[2]. But considering the enormous geometrical power of increase in every organism and as ⟨?⟩ every country, in ordinary cases ⟨countries⟩ must be stocked to full extent, reflection will show that this is the case. Malthus on man,—in animals no moral [check] restraint ⟨?⟩—they breed in time of year when provision most abundant, or season most favourable, every country has its seasons,—calculate robins—oscillating from years of destruction[3]. If proof were wanted let any singular change of climate ⟨occur⟩ here ⟨?⟩, how astoundingly some tribes ⟨?⟩ increase, also introduced animals[4], the

[1] Note in the original. "Good place to introduce, saying reasons hereafter to be given, how far I extend theory, say to all mammalia—reasons growing weaker and weaker."

[2] See *Origin*, Ed. i. pp. 62, 63, vi. p. 77, where similar reference is made to De Candolle; for Malthus see *Origin*, p. 5.

[3] This may possibly refer to the amount of destruction going on. See *Origin*, Ed. i. p. 68, vi. p. 84, where there is an estimate of a later date as to death-rate of birds in winter. "Calculate robins" probably refers to a calculation of the rate of increase of birds under favourable conditions.

[4] In the *Origin*, Ed. i. pp. 64, 65, vi. p. 80, he instances cattle and horses and certain plants in S. America and American species of plants in India, and further on, as unexpected effects of changed conditions, the enclosure of a heath, and the relation between the fertilisation of clover and the presence of cats (*Origin*, Ed. i. p. 74, vi. p. 91).

pressure is always ready,—capacity of alpine plants to endure other climates,—think of endless seeds scattered abroad,—forests regaining their percentage[1],—a thousand wedges[2] are being forced into the œconomy of nature. This requires much reflection; study Malthus and calculate rates of increase and remember the resistance,—only periodical.

The unavoidable effect of this ⟨is⟩ that many of every species are destroyed either in egg or [young or mature (the former state the more common)]. In the course of a thousand generations infinitesimally small differences must inevitably tell[3]; when unusually cold winter, or hot or dry summer comes, then out of the whole body of individuals of any species, if there be the smallest differences in their structure, habits, instincts [senses], health &c., ⟨it⟩ will on an average tell; as conditions change a rather larger proportion will be preserved: so if the chief check to increase falls on seeds or eggs, so will, in the course of 1000 generations or ten thousand, those seeds (like one with down to fly[4]) which fly furthest and get scattered most ultimately rear most plants, and such small differences tend to be hereditary like shades of expression in human countenance. So if one parent ⟨?⟩ fish deposits its egg in infinitesimally different circumstances, as in rather shallower or deeper water &c., it will then ⟨?⟩ tell.

Let hares[5] increase very slowly from change of climate affecting peculiar plants, and some other ⟨illegible⟩ rabbit decrease in same proportion [let this unsettle organisation of], a canine animal, who

[1] *Origin*, Ed. i. p. 74, vi. p. 91. "It has been observed that the trees now growing on...ancient Indian mounds...display the same beautiful diversity and proportion of kinds as in the surrounding virgin forests."

[2] The simile of the wedge occurs in the *Origin*, Ed. i. p. 67; it is deleted in Darwin's copy of the first edition: it does not occur in Ed. vi.

[3] In a rough summary at the close of the Essay, occur the words:—"Every creature lives by a struggle, smallest grain in balance must tell."

[4] Cf. *Origin*, Ed. i. p. 77, vi. p. 94.

[5] This is a repetition of what is given at p. 6.

formerly derived its chief sustenance by springing on rabbits or running them by scent, must decrease too and might thus readily become exterminated. But if its form varied very slightly, the long legged fleet ones, during a thousand years being selected, and the less fleet rigidly destroyed must, if no law of nature be opposed to it, alter forms.

Remember how soon Bakewell on the same principle altered cattle and Western, sheep,—carefully avoiding a cross (pigeons) with any breed. We cannot suppose that one plant tends to vary in fruit and another in flower, and another in flower and foliage,—some have been selected for both fruit and flower: that one animal varies in its covering and another not,—another in its milk. Take any organism and ask what is it useful for and on that point it will be found to vary,—cabbages in their leaf,—corn in size ⟨and⟩ quality of grain, both in times of year,—kidney beans for young pod and cotton for envelope of seeds &c. &c.: dogs in intellect, courage, fleetness and smell ⟨?⟩: pigeons in peculiarities approaching to monsters. This requires consideration,—should be introduced in first chapter if it holds, I believe it does. It is hypothetical at best[1].

Nature's variation far less, but such selection far more rigid and scrutinising. Man's races not [even so well] only not better adapted to conditions than other races, but often not ⟨?⟩ one race adapted to its conditions, as man keeps and propagates some alpine plants in garden. Nature lets ⟨an⟩ animal live, till on actual proof it is found less able to do the required work to serve the desired end, man judges solely by his eye, and knows not whether

[1] Compare *Origin*, Ed. i. p. 41, vi. p. 47. "I have seen it gravely remarked, that it was most fortunate that the strawberry began to vary just when gardeners began to attend closely to this plant. No doubt the strawberry had always varied since it was cultivated, but the slight varieties had been neglected."

nerves, muscles, arteries, are developed in proportion to the change of external form.

Besides selection by death, in bisexual animals ⟨illegible⟩ the selection in time of fullest vigour, namely struggle of males; even in animals which pair there seems a surplus ⟨?⟩ and a battle, possibly as in man more males produced than females, struggle of war or charms[1]. Hence that male which at that time is in fullest vigour, or best armed with arms or ornaments of its species, will gain in hundreds of generations some small advantage and transmit such characters to its offspring. So in female rearing its young, the most vigorous and skilful and industrious, ⟨whose⟩ instincts ⟨are⟩ best developed, will rear more young, probably possessing her good qualities, and a greater number will thus ⟨be⟩ prepared for the struggle of nature. Compared to man using a male alone of good breed. This latter section only of limited application, applies to variation of [specific] sexual characters. Introduce here contrast with Lamarck —absurdity of habit, or chance ?? or external conditions, making a woodpecker adapted to tree[2].

Before considering difficulties of theory of selection let us consider character of the races produced, as now explained, by nature. Conditions have varied slowly and the organisms best adapted in their whole course of life to the changed conditions have always been selected,—man selects small dog and afterwards gives it profusion of food,—selects a long-backed and short-legged breed and gives it no particular exercise to suit this function &c. &c. In ordinary cases nature has not allowed her race to

[1] Here we have the two types of sexual selection discussed in the *Origin*, Ed. i. pp. 88 et seq., vi. pp. 108 et seq.

[2] It is not obvious why the author objects to "chance" or "external conditions making a woodpecker." He allows that variation is ultimately referable to conditions and that the nature of the connexion is unknown, i.e. that the result is fortuitous. It is not clear in the original to how much of the passage the two ? refer.

be contaminated with a cross of another race, and agriculturists know how difficult they find always to prevent this,—effect would be trueness. This character and sterility when crossed, and generally a greater amount of difference, are two main features, which distinguish domestic races from species.

[Sterility not universal admitted by all[1]. *Gladiolus, Crinum, Calceolaria*[2] must be species if there be such a thing. Races of dogs and oxen: but certainly very general; indeed a gradation of sterility most perfect[3] very general. Some nearest species will not cross (crocus, some heath ⟨?⟩), some genera cross readily (fowls[4] and grouse, peacock &c.). Hybrids no ways monstrous quite perfect except secretions[5] hence even the mule has bred,—character of sterility, especially a few years ago ⟨?⟩ thought very much more universal than it now is, has been thought the distinguishing character; indeed it is obvious if all forms freely crossed, nature would be a chaos. But the very gradation of the character, even if it always existed in some degree which it does not, renders it impossible as marks ⟨?⟩ those ⟨?⟩ suppose distinct as species[6]]. Will analogy throw any light

[1] The meaning is "That sterility is not universal is admitted by all."

[2] See *Var. under Dom.*, Ed. 2, i. p. 388, where the garden forms of *Gladiolus* and *Calceolaria* are said to be derived from crosses between distinct species. Herbert's hybrid *Crinums* are discussed in the *Origin*, Ed. i. p. 250, vi. p. 370. It is well known that the author believed in a multiple origin of domestic dogs.

[3] The argument from gradation in sterility is given in the *Origin*, Ed. i. pp. 248, 255, vi. pp. 368, 375. In the *Origin*, I have not come across the cases mentioned, viz. crocus, heath, or grouse and fowl or peacock. For sterility between closely allied species, see *Origin*, Ed. i. p. 257, vi. p. 377. In the present essay the author does not distinguish between fertility between species and the fertility of the hybrid offspring, a point on which he insists in the *Origin*, Ed. i. p. 245, vi. p. 365.

[4] Ackermann (*Ber. d. Vereins f. Naturkunde zu Kassel*, 1898, p. 23) quotes from Gloger that a cross has been effected between a domestic hen and a *Tetrao tetrix*; the offspring died when three days old.

[5] No doubt the sexual cells are meant. I do not know on what evidence it is stated that the mule has bred.

[6] The sentence is all but illegible. I think that the author refers to forms usually ranked as varieties having been marked as species when it was

on the fact of the supposed races of nature being sterile, though none of the domestic ones are? Mr Herbert ⟨and⟩ Koelreuter have shown external differences will not guide one in knowing whether hybrids will be fertile or not, but the chief circumstance is constitutional differences[1], such as being adapted to different climate or soil, differences which [must] probably affect the whole body of the organism and not any one part. Now wild animals, taken out of their natural conditions, seldom breed. I do not refer to shows or to Zoological Societies where many animals unite, but ⟨do not?⟩ breed, and others will never unite, but to wild animals caught and kept *quite tame* left loose and well fed about houses and living many years. Hybrids produced almost as readily as pure breds. St Hilaire great distinction of tame and domestic,—elephants,—ferrets[2]. Reproductive organs not subject to disease in Zoological Garden. Dissection and microscope show that hybrid is in exactly same condition as another animal in the intervals of breeding season, or those animals which taken wild and *not bred* in domesticity, remain without breeding their whole lives. It should be observed that so far from domesticity being unfavourable in itself ⟨it⟩ makes more fertile: [when animal is domesticated and breeds, productive power increased from more food and selection of fertile races]. As far as animals go might be thought ⟨an⟩ effect on their mind and a special case.

But turning to plants we find same class of facts. I do not refer to seeds not ripening, perhaps the com-

found that they were sterile together. See the case of the red and blue *Anagallis* given from Gärtner in the *Origin*, Ed. i. p. 247, vi. p. 368.

[1] In the *Origin*, Ed. i. p. 258, where the author speaks of constitutional differences in this connexion, he specifies that they are confined to the reproductive system.

[2] The sensitiveness of the reproductive system to changed conditions is insisted on in the *Origin*, Ed. i. p. 8, vi. p. 10.

The ferret is mentioned, as being prolific in captivity, in *Var. under Dom.*, Ed. 2, ii. p. 90.

monest cause, but to plants not setting, which either is owing to some imperfection of ovule or pollen. Lindley says sterility is the [curse] bane of all propagators,—Linnæus about alpine plants. American bog plants,—pollen in exactly same state as in hybrids,—same in geraniums. Persian and Chinese[1] lilac will not seed in Italy and England. Probably double plants and all fruits owe their developed parts primarily ⟨?⟩ to sterility and extra food thus ⟨?⟩ applied[2]. There is here gradation ⟨in⟩ sterility and then parts, like diseases, are transmitted hereditarily. We cannot assign any cause why the Pontic Azalea produces plenty of pollen and not American[3], why common lilac seeds and not Persian, we see no difference in healthiness. We know not on what circumstances these facts depend, why ferret breeds, and cheetah[4], elephant and pig in India will not.

Now in crossing it is certain every peculiarity in form and constitution is transmitted: an alpine plant transmits its alpine tendency to its offspring, an American plant its American-bog constitution, and ⟨with⟩ animals, those peculiarities, on which[5] when placed out of their natural conditions they are incapable of breeding; and moreover they transmit every part of their constitution, their

[1] Lindley's remark is quoted in the *Origin*, Ed. i. p. 9. Linnæus' remark is to the effect that Alpine plants tend to be sterile under cultivation (see *Var. under Dom.*, Ed. 2, ii. p. 147). In the same place the author speaks of peat-loving plants being sterile in our gardens,—no doubt the American bog-plants referred to above. On the following page (p. 148) the sterility of the lilac (*Syringa persica* and *chinensis*) is referred to.

[2] The author probably means that the increase in the petals is due to a greater food supply being available for them owing to sterility. See the discussion in *Var. under Dom.*, Ed. 2, ii. p. 151. It must be noted that doubleness of the flower may exist without noticeable sterility.

[3] I have not come across this case in the author's works.

[4] For the somewhat doubtful case of the cheetah (*Felis jubata*) see *Var. under Dom.*, Ed. 2, ii. p. 133. I do not know to what fact "pig in India" refers.

[5] This sentence should run "on which depends their incapacity to breed in unnatural conditions."

respiration, their pulse, their instinct, which are all suddenly modified, can it be wondered at that they are incapable of breeding? I think it may be truly said it would be more wonderful if they did. But it may be asked why have not the recognised varieties, supposed to have been produced through the means of man, [not refused to breed] have all bred[1]. Variation depends on change of condition and selection[2], as far as man's systematic or unsystematic selection ⟨has⟩ gone; he takes external form, has little power from ignorance over internal invisible constitutional differences. Races which have long been domesticated, and have much varied, are precisely those which were capable of bearing great changes, whose constitutions were adapted to a diversity of climates. Nature changes slowly and by degrees. According to many authors probably breeds of dogs are another case of modified species freely crossing. There is no variety which ⟨illegible⟩ has been ⟨illegible⟩ adapted to peculiar soil or situation for a thousand years and another rigorously adapted to another, till such can be produced, the question is not tried[3]. Man in past ages, could transport into different climates, animals and plants which would freely propagate in such new climates. Nature could effect, with selection, such changes slowly, so that precisely those animals which are adapted to submit to great changes have given rise to diverse races,—and indeed great doubt on this head[4].

[1] This sentence ends in confusion: it should clearly close with the words "refused to breed" in place of the bracket and the present concluding phrase.

[2] The author doubtless refers to the change produced by the *summation* of variation by means of selection.

[3] The meaning of this sentence is made clear by a passage in the MS. of 1844:—"Until man selects two varieties from the same stock, adapted to two climates or to other different external conditions, and confines each rigidly for one or several thousand years to such conditions, always selecting the individuals best adapted to them, he cannot be said to have even commenced the experiment." That is, the attempt to produce mutually sterile domestic breeds.

[4] This passage is to some extent a repetition of a previous one and may

Before leaving this subject well to observe that it was shown that a certain amount of variation is consequent on mere act of reproduction, both by buds and sexually,—is vastly increased when parents exposed for some generations to new conditions[1], and we now find that many animals when exposed for first time to very new conditions, are ⟨as⟩ incapable of breeding as hybrids. It [probably] bears also on supposed fact of crossed animals when not infertile, as in mongrels, tending to vary much, as likewise seems to be the case, when true hybrids possess just sufficient fertility to propagate with the parent breeds and *inter se* for some generations. This is Koelreuter's belief. These facts throw light on each other and support the truth of each other, we see throughout a connection between the reproductive faculties and exposure to changed conditions of life whether by crossing or exposure of the individuals[2].

Difficulties on theory of selection[3]. It may be objected such perfect organs as eye and ear, could never be formed, in latter less difficulty as gradations more perfect; at first appears monstrous and to ⟨the⟩ end appears difficulty. But think of gradation, even now manifest, (Tibia and Fibula). Everyone will allow if every fossil preserved, gradation

have been intended to replace an earlier sentence. I have thought it best to give both. In the *Origin*, Ed. i. p. 141, vi. p. 176, the author gives his opinion that the power of resisting diverse conditions, seen in man and his domestic animals, is an example "of a very common flexibility of constitution."

1 In the *Origin*, Ed. i. Chs. I. and v., the author does not admit reproduction, apart from environment, as being a cause of variation. With regard to the cumulative effect of new conditions there are many passages in the *Origin*, Ed. i. e.g. pp. 7, 12, vi. pp. 8, 14.

2 As already pointed out, this is the important principle investigated in the author's *Cross and Self-Fertilisation*. Professor Bateson has suggested to me that the experiments should be repeated with gametically pure individuals.

3 In the *Origin* a chapter is given up to "difficulties on theory": the discussion in the present essay seems slight even when it is remembered how small a space is here available. For *Tibia* &c. see p. 48.

infinitely more perfect; for possibility of selection a perfect (?) gradation is required. Different groups of structure, slight gradation in each group,—every analogy renders it probable that intermediate forms have existed. Be it remembered what strange metamorphoses; part of eye, not directly connected with vision, might come to be [thus used] gradually worked in for this end,—swimming bladder by gradation of structure is admitted to belong to the ear system, —rattlesnake. [Woodpecker best adapted to climb.] In some cases gradation not possible,—as vertebræ, —actually vary in domestic animals,—less difficult if growth followed. Looking to whole animals, a bat formed not for flight[1]. Suppose we had flying fish[2] and not one of our now called flying fish preserved, who would have guessed intermediate habits. Woodpeckers and tree-frogs both live in countries where no trees[3].

The gradations by which each individual organ has arrived at its present state, and each individual animal with its aggregate of organs has arrived, probably never could be known, and all present great difficulties. I merely wish to show that the proposition is not so monstrous as it at first appears, and that if good reason can be advanced for believing the species have descended from common parents, the difficulty of imagining intermediate forms of structure not sufficient to make one at once reject the theory.

[1] This may be interpreted "The general structure of a bat is the same as that of non-flying mammals."

[2] That is truly winged fish.

[3] The terrestrial woodpecker of S. America formed the subject of a paper by Darwin, *Proc. Zool. Soc.*, 1870. See *Life and Letters*, vol. iii. p. 153.

§ III. ⟨On Variation in instincts and other mental attributes.⟩

The mental powers of different animals in wild and tame state [present still greater difficulties] require a separate section. Be it remembered I have nothing to do with origin of memory, attention, and the different faculties of the mind[1], but merely with their differences in each of the great divisions of nature. Disposition, courage, pertinacity ⟨?⟩, suspicion, restlessness, ill-temper, sagacity and ⟨the⟩ reverse unquestionably vary in animals and are inherited (Cuba wildness dogs, rabbits, fear against particular object as man Galapagos[2]). Habits purely corporeal, breeding season &c., time of going to rest &c., vary and are hereditary, like the analogous habits of plants which vary and are inherited. Habits of body, as manner of movement d°. and d°. Habits, as pointing and setting on certain occasions d°. Taste for hunting certain objects and manner of doing so—sheep-dog. These are shown clearly by crossing and their analogy with true instinct thus shown,—retriever. Do not know objects for which they do it. Lord Brougham's definition[3]. Origin partly habit, but the amount necessarily unknown, partly selection. Young pointers pointing stones and sheep—tumbling pigeons—sheep[4] going back to place where born.

[1] The same proviso occurs in the *Origin*, Ed. i. p. 207, vi. p. 319.

[2] The tameness of the birds in the Galapagos is described in the *Journal of Researches* (1860), p. 398. Dogs and rabbits are probably mentioned as cases in which the hereditary fear of man has been lost. In the 1844 MS. the author states that the Cuban feral dog shows great natural wildness, even when caught quite young.

[3] In the *Origin*, Ed. i. p. 207, vi. p. 319, he refuses to define instinct. For Lord Brougham's definition see his *Dissertations on Subjects of Science etc.*, 1839, p. 27.

[4] See James Hogg (the Ettrick Shepherd), Works, 1865, *Tales and Sketches*, p. 403.

Instinct aided by reason, as in the taylor-bird[1]. Taught by parents, cows choosing food, birds singing. Instincts vary in wild state (birds get wilder) often lost[2]; more perfect,—nest without roof. These facts [only clear way] show how incomprehensibly brain has power of transmitting intellectual operations.

Faculties[3] distinct from true instincts,—finding [way]. It must I think be admitted that habits whether congenital or acquired by practice [sometimes] often become inherited[4]; instincts, influence, equally with structure, the preservation of animals; therefore selection must, with changing conditions tend to modify the inherited habits of animals. If this be admitted it will be found *possible* that many of the strangest instincts may be thus acquired. I may observe, without attempting definition, that an inherited habit or trick (trick because may be born) fulfils closely what we mean by instinct. A habit is often performed unconsciously, the strangest habits become associated, d°. tricks, going in certain spots &c. &c., even against will, is excited by external agencies, and looks not to the end,—a person playing a pianoforte. If such a habit were transmitted it would make a marvellous instinct. Let us consider some of the most difficult cases of instincts, whether they could be *possibly* acquired. I do not say *probably*, for that belongs to our 3rd Part[5], I beg this may be remembered, nor do I mean to attempt to show exact method. I want only to show that

[1] This refers to the tailor-bird making use of manufactured thread supplied to it, instead of thread twisted by itself.

[2] *Often lost* applies to *instinct*: *birds get wilder* is printed in a parenthesis because it was apparently added as an after-thought. *Nest without roof* refers to the water-ousel omitting to vault its nest when building in a protected situation.

[3] In the MS. of 1844 is an interesting discussion on *faculty* as distinct from *instinct*.

[4] At this date and for long afterwards the inheritance of acquired characters was assumed to occur.

[5] Part II. is here intended: see the Introduction.

whole theory ought not at once to be rejected on this score.

Every instinct must, by my theory, have been acquired gradually by slight changes ⟨illegible⟩ of former instinct, each change being useful to its then species. Shamming death struck me at first as remarkable objection. I found none really sham death[1], and that there is gradation; now no one doubts that those insects which do it either more or less, do it for some good, if then any species was led to do it more, and then ⟨?⟩ escaped &c. &c.

Take migratory instincts, faculty distinct from instinct, animals have notion of time,—like savages. Ordinary finding way by memory, but how does savage find way across country,—as incomprehensible to us, as animal to them,—geological changes,—fishes in river,—case of sheep in Spain[2]. Architectural instincts,—a manufacturer's employee in making single articles extraordinary skill,—often said seem to make it almost ⟨illegible⟩, child born with such a notion of playing[3],—we can fancy tailoring acquired in same perfection,—mixture of reason,—water-ouzel,—taylor-bird,—gradation of simple nest to most complicated.

Bees again, distinction of faculty,—how they make a hexagon,—Waterhouse's theory[4],—the impulse to use whatever faculty they possess,—the taylor-bird has the faculty of sewing with beak, instinct impels him to do it.

Last case of parent feeding young with different food (take case of Galapagos birds, gradation from

[1] The meaning is that the attitude assumed in *shamming* is not accurately like that of death.

[2] This refers to the *transandantes* sheep mentioned in the MS. of 1844, as having acquired a migratory instinct.

[3] In the *Origin*, Ed. i. p. 209, vi. p. 321, Mozart's pseudo-instinctive skill in piano-playing is mentioned. See *Phil. Trans.*, 1770, p. 54.

[4] In the discussion on bees' cells, *Origin*, Ed. i. p. 225, vi. p. 343, the author acknowledges that his theory originated in Waterhouse's observations.

Hawfinch to Sylvia) selection and habit might lead old birds to vary taste ⟨?⟩ and form, leaving their instinct of feeding their young with same food[1], —or I see no difficulty in parents being forced or induced to vary the food brought, and selection adapting the young ones to it, and thus by degree any amount of diversity might be arrived at. Although we can never hope to see the course revealed by which different instincts have been acquired, for we have only present animals (not well known) to judge of the course of gradation, yet once grant the principle of habits, whether congenital or acquired by experience, being inherited and I can see no limit to the [amount of variation] extraordinariness ⟨?⟩ of the habits thus acquired.

Summing up this Division. If variation be admitted to occur occasionally in some wild animals, and how can we doubt it, when we see [all] thousands ⟨of⟩ organisms, for whatever use taken by man, do vary. If we admit such variations tend to be hereditary, and how can we doubt it when we ⟨remember⟩ resemblances of features and character, —disease and monstrosities inherited and endless races produced (1200 cabbages). If we admit selection is steadily at work, and who will doubt it, when he considers amount of food on an average fixed and reproductive powers act in geometrical ratio. If we admit that external conditions vary, as all geology proclaims, they have done and are now doing, —then, if no law of nature be opposed, there must occasionally be formed races, [slightly] differing from the parent races. So then any such law[2], none is

[1] The hawfinch- and *Sylvia*-types are figured in the *Journal of Researches*, p. 379. The discussion of change of form in relation to change of instinct is not clear, and I find it impossible to suggest a paraphrase.

[2] I should interpret this obscure sentence as follows, "No such opposing law is known, but in all works on the subject a law is (in flat contradiction to all known facts) assumed to limit the possible amount of variation." In the *Origin*, the author never limits the power of variation, as far as I know.

known, but in all works it is assumed, in ⟨?⟩ flat contradiction to all known facts, that the amount of possible variation is soon acquired. Are not all the most varied species, the oldest domesticated: who ⟨would⟩ think that horses or corn could be produced? Take dahlia and potato, who will pretend in 5000 years[1] ⟨that great changes might not be effected⟩: perfectly adapted to conditions and then again brought into varying conditions. Think what has been done in few last years, look at pigeons, and cattle. With the amount of food man can produce he may have arrived at limit of fatness or size, or thickness of wool ⟨?⟩, but these are the most trivial points, but even in these I conclude it is impossible to say we know the limit of variation. And therefore with the [adapting] selecting power of nature, infinitely wise compared to those of man, ⟨I conclude⟩ that it is impossible to say we know the limit of races, which would be true ⟨to their⟩ kind; if of different constitutions would probably be infertile one with another, and which might be adapted in the most singular and admirable manner, according to their wants, to external nature and to other surrounding organisms,—such races would be species. But is there any evidence ⟨that⟩ species ⟨have⟩ been thus produced, this is a question wholly independent of all previous points, and which on examination of the kingdom of nature ⟨we⟩ ought to answer one way or another.

[1] In *Var. under Dom.* Ed. 2, ii. p. 263, the *Dahlia* is described as showing sensitiveness to conditions in 1841. All the varieties of the *Dahlia* are said to have arisen since 1804 (*ibid.* i. p. 393).

PART II[1].

§§ IV. & V. ⟨On the evidence from Geology.⟩

I may premise, that according to the view ordinarily received, the myriads of organisms peopling this world have been created by so many distinct acts of creation. As we know nothing of the ⟨illegible⟩ will of a Creator,—we can see no reason why there should exist any relation between the organisms thus created; or again, they might be created according to any scheme. But it would be marvellous if this scheme should be the same as would result from the descent of groups of organisms from [certain] the same parents, according to the circumstances, just attempted to be developed.

With equal probability did old cosmogonists say fossils were created, as we now see them, with a false resemblance to living beings[2]; what would the Astronomer say to the doctrine that the planets moved ⟨not⟩ according to the law of gravitation, but from the Creator having willed each separate planet to move in its particular orbit? I believe such a proposition (if we remove all prejudices) would be as legitimate as to admit that certain groups of living and extinct organisms, in their distribution, in their structure and in their relations one to another and to external conditions, agreed with the theory

[1] In the original MS. the heading is: Part III.; but Part II. is clearly intended; for details see the Introduction. I have not been able to discover where § IV. ends and § V. begins.

[2] This passage corresponds roughly to the conclusion of the *Origin*, see Ed. i. p. 482, vi. p. 661.

and showed signs of common descent, and yet were created distinct. As long as it was thought impossible that organisms should vary, or should anyhow become adapted to other organisms in a complicated manner, and yet be separated from them by an impassable barrier of sterility[1], it was justifiable, even with some appearance in favour of a common descent, to admit distinct creation according to the will of an Omniscient Creator; or, for it is the same thing, to say with Whewell that the beginnings of all things surpass the comprehension of man. In the former sections I have endeavoured to show that such variation or specification is not impossible, nay, in many points of view is absolutely probable. What then is the evidence in favour of it and what the evidence against it. With our imperfect knowledge of past ages [surely there will be some] it would be strange if the imperfection did not create some unfavourable evidence.

Give sketch of the Past,—beginning with facts appearing hostile under present knowledge,—then proceed to geograph. distribution,—order of appearance,—affinities,—morphology &c., &c.

Our theory requires a very gradual introduction of new forms[2], and extermination of the old (to which we shall revert). The extermination of old may sometimes be rapid, but never the introduction. In the groups descended from common parent, our theory requires a perfect gradation not differing more than breed⟨s⟩ of cattle, or potatoes, or cabbages in forms. I do not mean that a graduated series of animals must have existed, intermediate between horse, mouse, tapir[3], elephant [or fowl and peacock],

[1] A similar passage occurs in the conclusion of the *Origin*, Ed. i. p. 481, vi. p. 659.

[2] See *Origin*, Ed. i. p. 312, vi. p. 453.

[3] See *Origin*, Ed. i. pp. 280, 281, vi. p. 414. The author uses his experience of pigeons for examples for what he means by *intermediate*; the instance of the horse and tapir also occurs.

but that these must have had a common parent, and between horse and this ⟨?⟩ parent &c., &c., but the common parent may possibly have differed more from either than the two do now from each other. Now what evidence of this is there? So perfect gradation in some departments, that some naturalists have thought that in some large divisions, if all existing forms were collected, a near approach to perfect gradation would be made. But such a notion is preposterous with respect to all, but evidently so with mammals. Other naturalists have thought this would be so if all the specimens entombed in the strata were collected[1]. I conceive there is no probability whatever of this; nevertheless it is certain all the numerous fossil forms fall in⟨to⟩, as Buckland remarks, *not* present classes, families and genera, they fall between them: so is it with new discoveries of existing forms. Most ancient fossils, that is most separated ⟨by⟩ space of time, are most apt to fall between the classes—(but organisms from those countries most separated by space also fall between the classes ⟨*e.g.*⟩ Ornithorhyncus?). As far as geological discoveries ⟨go⟩ they tend towards such gradation[2]. Illustrate it with net. Toxodon,—tibia and fibula,—dog and otter,—but so utterly improbable is ⟨it⟩, in *ex. gr.* Pachydermata, to compose series as perfect as cattle, that if, as many geologists seem to

[1] The absence of intermediate forms between living organisms (and also as regards fossils) is discussed in the *Origin*, Ed. i. pp. 279, 280, vi. p. 413. In the above discussion there is no evidence that the author felt this difficulty so strongly as it is expressed in the *Origin*, Ed. i. p. 299,—as perhaps "the most obvious and gravest objection that can be urged against my theory." But in a rough summary written on the back of the penultimate page of the MS. he refers to the geological evidence:—"Evidence, as far as it does go, is favourable, exceedingly incomplete,—greatest difficulty on this theory. I am convinced not insuperable." Buckland's remarks are given in the *Origin*, Ed. i. p. 329, vi. p. 471.

[2] That the evidence of geology, as far as it goes, is favourable to the theory of descent is claimed in the *Origin*, Ed. i. pp. 343—345, vi. pp. 490—492. For the reference to *net* in the following sentence, see Note 1, p. 48, of this Essay.

infer, each separate formation presents even an approach to a consecutive history, my theory must be given up. Even if it were consecutive, it would only collect series of one district in our present state of knowledge; but what probability is there that any one formation during the *immense* period which has elapsed during each period will *generally* present a consecutive history. [Compare number living at one period to fossils preserved—look at enormous periods of time.]

Referring only to marine animals, which are obviously most likely to be preserved, they must live where ⟨?⟩ sediment (of a kind favourable for preservation, not sand and pebble)[1] is depositing quickly and over large area and must be thickly capped, ⟨illegible⟩ littoral deposits: for otherwise denudation ⟨will destroy them⟩,—they must live in a shallow space which sediment will tend to fill up,—as movement is ⟨in ?⟩ progress if soon brought ⟨?⟩ up ⟨?⟩ subject to denudation,—[if] as during subsidence favourable, accords with facts of European deposits[2], but subsidence apt to destroy agents which produce sediment[3].

I believe safely inferred ⟨that⟩ groups of marine ⟨?⟩ fossils only preserved for future ages where sediment goes on long ⟨and⟩ continuous⟨ly⟩ and with rapid but not too rapid deposition in ⟨an⟩ area of subsidence. In how few places in any one region like Europe will ⟨?⟩ these contingencies be going on ? Hence ⟨?⟩ in

[1] See *Origin*, Ed. i. p. 288, vi. p. 422. "The remains that do become embedded, if in sand and gravel, will, when the beds are upraised, generally be dissolved by the percolation of rain-water."

[2] The position of the following is not clear:—"Think of immense differences in nature of European deposits,—without interposing new causes,—think of time required by present slow changes, to cause, on very same area, such diverse deposits, iron-sand, chalk, sand, coral, clay!"

[3] The paragraph which ends here is difficult to interpret. In spite of obscurity it is easy to recognize the general resemblance to the discussion on the importance of subsidence given in the *Origin*, Ed. i. pp. 290 et seq., vi. pp. 422 et seq.

past ages mere [gaps] pages preserved[1]. Lyell's doctrine carried to extreme,—we shall understand difficulty if it be asked:—what chance of series of gradation between cattle by ⟨illegible⟩ at age ⟨illegible⟩ as far back as Miocene[2]? We know then cattle existed. Compare number of living,—immense duration of each period,—fewness of fossils.

This only refers to consecutiveness of history of organisms of each formation.

The foregoing argument will show firstly, that formations are distinct merely from want of fossils ⟨of intermediate beds⟩, and secondly, that each formation is full of gaps, has been advanced to account for *fewness* of *preserved* organisms compared to what have lived on the world. The very same argument explains why in older formations the organisms appear to come on and disappear suddenly,—but in [later] tertiary not quite suddenly[3], in later tertiary gradually,—becoming rare and disappearing,—some have disappeared within man's time. It is obvious that our theory requires gradual and nearly uniform introduction, possibly more sudden extermination,—subsidence of continent of Australia &c., &c.

Our theory requires that the first form which existed of each of the great divisions would present points intermediate between existing ones, but immensely different. Most geologists believe Silurian[4] fossils are those which first existed in the whole world,

[1] See Note 3, p. 27.

[2] Compare *Origin*, Ed. i. p. 298, vi. p. 437. "We shall, perhaps, best perceive the improbability of our being enabled to connect species by numerous, fine, intermediate, fossil links, by asking ourselves whether, for instance, geologists at some future period will be able to prove that our different breeds of cattle, sheep, horses, and dogs have descended from a single stock or from several aboriginal stocks."

[3] The sudden appearance of groups of allied species in the lowest known fossiliferous strata is discussed in the *Origin*, Ed. i. p. 306, vi. p. 446. The gradual appearance in the later strata occurs in the *Origin*, Ed. i. p. 312, vi. p. 453.

[4] Compare *Origin*, Ed. i. p. 307, vi. p. 448.

not those which have chanced to be the oldest not destroyed,—or the first which existed in profoundly deep seas in progress of conversion from sea to land: if they are first they ⟨? we⟩ give up. Not so Hutton or Lyell: if first reptile[1] of Red Sandstone ⟨?⟩ really was first which existed: if Pachyderm[2] of Paris was first which existed: fish of Devonian: dragon fly of Lias: for we cannot suppose them the progenitors: they agree too closely with existing divisions. But geologists consider Europe as ⟨?⟩ a passage from sea to island ⟨?⟩ to continent (except Wealden, see Lyell). These animals therefore, I consider then mere introduction ⟨?⟩ from continents long since submerged.

Finally, if views of some geologists be correct, my theory must be given up. [Lyell's views, as far as they go, are in *favour*, but they go so little in favour, and so much more is required, that it may ⟨be⟩ viewed as objection.] If geology present us with mere pages in chapters, towards end of ⟨a⟩ history, formed by tearing out bundles of leaves, and each page illustrating merely a small portion of the organisms of that time, the facts accord perfectly with my theory[3].

[1] I have interpreted as *Sandstone* a scrawl which I first read as *Sea*; I have done so at the suggestion of Professor Judd, who points out that "footprints in the red sandstone were known at that time, and geologists were not then particular to distinguish between Amphibians and Reptiles."

[2] This refers to Cuvier's discovery of *Palæotherium* &c. at Montmartre.

[3] This simile is more fully given in the *Origin*, Ed. i. p. 310, vi. p. 452. "For my part, following out Lyell's metaphor, I look at the natural geological record, as a history of the world imperfectly kept, and written in a changing dialect; of this history we possess the last volume alone, relating only to two or three countries. Of this volume, only here and there a short chapter has been preserved; and of each page, only here and there a few lines. Each word of the slowly-changing language, in which the history is supposed to be written, being more or less different in the interrupted succession of chapters, may represent the apparently abruptly changed forms of life, entombed in our consecutive, but widely separated formations." Professor Judd has been good enough to point out to me, that Darwin's metaphor is founded on the comparison of geology to history in Ch. i. of the *Principles of Geology*, Ed. i. 1830, vol. i. pp. 1—4. Professor Judd has also called my attention to another passage,—*Principles*, Ed. i. 1833, vol. iii. p. 33, when Lyell imagines an historian examining "two buried cities at the foot of Vesuvius, immediately superimposed upon each

Extermination. We have seen that in later periods the organisms have disappeared by degrees and [perhaps] probably by degrees in earlier, and I have said our theory requires it. As many naturalists seem to think extermination a most mysterious circumstance[1] and call in astonishing agencies, it is well to recall what we have shown concerning the struggle of nature. An exterminating agency is at work with every organism: we scarcely see it: if robins would increase to thousands in ten years how severe must the process be. How imperceptible a small increase: fossils become rare: possibly sudden extermination as Australia, but as present means very slow and many means of escape, I shall doubt very sudden exterminations. Who can explain why some species abound more,—why does marsh titmouse, or ring-ouzel, now little change,—why is one sea-slug rare and another common on our coasts,—why one species of Rhinoceros more than another,—why is ⟨illegible⟩ tiger of India so rare? Curious and general sources of error, the place of an organism is instantly filled up.

We know state of earth has changed, and as earthquakes and tides go on, the state must change,—many geologists believe a slow gradual cooling. Now let us see in accordance with principles of [variation] specification explained in Sect. II. how species would probably be introduced and how such results accord with what is known.

other." The historian would discover that the inhabitants of the lower town were Greeks while those of the upper one were Italians. But he would be wrong in supposing that there had been a sudden change from the Greek to the Italian language in Campania. I think it is clear that Darwin's metaphor is partly taken from this passage. See for instance (in the above passage from the *Origin*) such phrases as "history...written in a changing dialect"—"apparently abruptly changed forms of life." The passage within [ ] in the above paragraph:—"Lyell's views as far as they go &c.," no doubt refers, as Professor Judd points out, to Lyell not going so far as Darwin on the question of the imperfection of the geological record.

[1] On rarity and extinction see *Origin*, Ed. i. pp. 109, 319, vi. pp. 133, 461.

The first fact geology proclaims is immense number of extinct forms, and new appearances. Tertiary strata leads to belief, that forms gradually become rare and disappear and are gradually supplied by others. We see some forms now becoming rare and disappearing, we know of no sudden creation: in older periods the forms *appear* to come in suddenly, scene shifts: but even here Devonian, Permian &c. [keep on supplying new links in chain]—Genera and higher forms come on and disappear, in same way leaving a species on one or more stages below that in which the form abounded.

⟨Geographical Distribution.⟩

§ vi. Let us consider the absolute state of distribution of organisms of earth's face.

Referring chiefly, but not exclusively (from difficulty of transport, fewness, and the distinct characteristics of groups) to Mammalia; and first considering the three or four main [regions] divisions; North America, Europe, Asia, including greater part of E. Indian Archipelago and Africa are intimately allied. Africa most distinct, especially most southern parts. And the Arctic regions, which unite N. America, Asia and Europe, only separated (if we travel one way by Behring's St.) by a narrow strait, is most intimately allied, indeed forms but one restricted group. Next comes S. America,—then Australia, Madagascar (and some small islands which stand very remote from the land). Looking at these main divisions separately, the organisms vary according to changes in condition[1] of different parts. But besides this, barriers of every kind seem to separate

[1] In the *Origin*, Ed. i. p. 346, vi. p. 493, the author begins his discussion on geographical distribution by minimising the effect of physical conditions. He lays great stress on the effect of *barriers*, as in the present Essay.

regions in a greater degree than proportionally to the difference of climates on each side. Thus great chains of mountains, spaces of sea between islands and continents, even great rivers and deserts. In fact the amount ⟨of⟩ difference in the organisms bears a certain, but not invariable relation to the amount of physical difficulties to transit[1].

There are some curious exceptions, namely, similarity of fauna of mountains of Europe and N. America and Lapland. Other cases just ⟨the⟩ reverse, mountains of eastern S. America, Altai (?), S. India (?)[2]: mountain summits of islands often eminently peculiar. Fauna generally of some islands, even when close, very dissimilar, in others very similar. [I am here led to observe one or more centres of creation[3].]

The simple geologist can explain many of the foregoing cases of distribution. Subsidence of a continent in which free means of dispersal, would drive the lowland plants up to the mountains, now converted into islands, and the semi-alpine plants would take place of alpine, and alpine be destroyed, if mountains originally were not of great height. So we may see, during gradual changes[4] of climate on a continent, the propagation of species would vary and adapt themselves to small changes

[1] Note in the original, "Would it be more striking if we took animals, take Rhinoceros, and study their habitats?"

[2] Note by Mr A. R. Wallace. "The want of similarity referred to, is, between the mountains of Brazil and Guiana and those of the Andes. Also those of the Indian peninsula as compared with the Himalayas. In both cases there is continuous intervening land.

"The islands referred to were, no doubt, the Galapagos for dissimilarity from S. America; our own Islands as compared with Europe, and perhaps Java, for similarity with continental Asia."

[3] The arguments against multiple centres of creation are given in the *Origin*, Ed. i. p. 352, vi. p. 499.

[4] In the *Origin*, Ed. i. p. 366, vi. p. 516, the author does not give his views on the distribution of alpine plants as original but refers to Edward Forbes' work (*Geolog. Survey Memoirs*, 1846). In his autobiography, Darwin refers to this. "I was forestalled" he says, "in only one important point, which my vanity has always made me regret." (*Life and Letters*, i. p. 88.)

causing much extermination[1]. The mountains of Europe were quite lately covered with ice, and the lowlands probably partaking of the Arctic climate and Fauna. Then as climate changed, arctic fauna would take place of ice, and an inundation of plants from different temperate countries ⟨would⟩ seize the lowlands, leaving islands of arctic forms. But if this had happened on an island, whence could the new forms have come,—here the geologist calls in creationists. If island formed, the geologist will suggest ⟨that⟩ many of the forms might have been borne from nearest land, but if peculiar, he calls in creationist,—as such island rises in height &c., he still more calls in creation. The creationist tells one, on a ⟨illegible⟩ spot the American spirit of creation makes *Orpheus* and *Tyrannus* and American doves, and in accordance with past and extinct forms, but no persistent relation between areas and distribution, Geologico-Geograph.-Distribution.

[1] ⟨The following is written on the back of a page of the MS.⟩ Discuss one or more centres of creation: allude strongly to facilities of dispersal and amount of geological change: allude to mountain-summits afterwards to be referred to. The distribution varies, as everyone knows, according to adaptation, explain going from N. to S. how we come to fresh groups of species in the same general region, but besides this we find difference, according to greatness of barriers, in greater proportion than can be well accounted for by adaptation. ⟨On representive species see *Origin*, Ed. i. p. 349, vi. p. 496.⟩ This very striking when we think of cattle of Pampas, plants ⟨?⟩ &c. &c. Then go into discussion; this holds with 3 or 4 main divisions as well as the endless minor ones in each of these 4 great ones: in these I chiefly refer to mammalia &c. &c. The similarity of type, but not in species, in same continent has been much less insisted on than the dissimilarity of different great regions generically: it is more striking.

⟨I have here omitted an incomprehensible sentence.⟩ Galapagos Islands, Tristan d'Acunha, *volcanic* islands covered with craters we know lately did not support any organisms. How unlike these islands in nature to neighbouring lands. These facts perhaps more striking than almost any others. [Geology apt to affect geography therefore we ought to expect to find the above.] Geological-geographical distribution. In looking to past times we find Australia equally distinct. S. America was distinct, though with more forms in common. N. America its nearest neighbour more in common,—in some respects more, in some less allied to Europe. Europe we find ⟨?⟩ equally European. For Europe is now part of Asia though not ⟨illegible⟩. Africa unknown,—examples, Elephant, Rhinoceros, Hippopotamus, Hyaena. As geology destroys geography we cannot be surprised in going far back we find Marsupials and Edentata in Europe: but geology destroys geography.

Now according to analogy of domesticated animals let us see what would result. Let us take case of farmer on Pampas, where everything approaches nearer to state of nature. He works on organisms having strong tendency to vary: and he knows ⟨that the⟩ only way to make a distinct breed is to select and separate. It would be useless to separate the best bulls and pair with best cows if their offspring run loose and bred with the other herds, and tendency to reversion not counteracted; he would endeavour therefore to get his cows on islands and then commence his work of selection. If several farmers in different *rincons*[1] were to set to work, especially if with different objects, several breeds would soon be produced. So would it be with horticulturist and so history of every plant shows; the number of varieties[2] increase in proportion to care bestowed on their selection and, with crossing plants, separation. Now, according to this analogy, change of external conditions, and isolation either by chance landing ⟨of⟩ a form on an island, or subsidence dividing a continent, or great chain of mountains, and the number of individuals not being numerous will best favour variation and selection[3]. No doubt change could be effected in same country without any barrier by long continued selection on one species: even in case of a plant not capable of crossing would easier get possession and solely

[1] *Rincon* in Spanish means a *nook* or *corner*, it is here probably used to mean a small farm.

[2] The following is written across the page: "No one would expect a set of similar varieties to be produced in the different countries, so species different."

[3] ⟨The following passage seems to have been meant to follow here.⟩ The parent of an organism, we may generally suppose to be in less favourable condition than the selected offspring and therefore generally in fewer numbers. (This is not borne out by horticulture, mere hypothesis; as an organism in favourable conditions might by selection be adapted to still more favourable conditions.)

Barrier would further act in preventing species formed in one part migrating to another part.

occupy an island[1]. Now we can at once see that ⟨if⟩ two parts of a continent isolated, new species thus generated in them, would have closest affinities, like cattle in counties of England: if barrier afterwards destroyed one species might destroy the other or both keep their ground. So if island formed near continent, let it be ever so different, that continent would supply inhabitants, and new species (like the old) would be allied with that continent. An island generally very different soil and climate, and number and order of inhabitants supplied by chance, no point so favourable for generation of new species[2],—especially the mountains, hence, so it is. As isolated mountains formed in a plain country (if such happens) is an island. As other islands formed, the old species would spread and thus extend and the fauna of distant island might ultimately meet and a continent formed between them. No one doubts continents formed by repeated elevations and depressions[3]. In looking backwards, but not so far that all geographical boundaries are destroyed, we can thus at once see why existing forms are related to the extinct in the same manner as existing ones are in some part of existing continent. By chance we might even have one or two absolute parent fossils.

The detection of transitional forms would be rendered more difficult on rising point of land.

The distribution therefore in the above enumer-

[1] ⟨The following notes occur on the back of the page.⟩ Number of species not related to capabilities of the country: furthermore not always those best adapted, perhaps explained by creationists by changes and progress. ⟨See p. 34, note 1.⟩

Although creationists can, by help of geology, explain much, how can he explain the marked relation of past and present in same area, the varying relation in other cases, between past and present, the relation of different parts of same great area. If island, to adjoining continent, if quite different, on mountain summits,—the number of individuals not being related to capabilities, or how &c.—our theory, I believe, can throw much light and all facts accord.

[2] See *Origin*, Ed. i. p. 390, vi. p. 543.

[3] On oscillation see *Origin*, Ed. i. p. 291, vi. p. 426.

ated points, even the trivial ones, which on any other ⟨theory ?⟩ can be viewed as so many ultimate facts, all follow ⟨in⟩ a simple manner on the theory of the occurrence of species by ⟨illegible⟩ and being adapted by selection to ⟨illegible⟩, conjoined with their power of dispersal, and the steady geographico-geological changes which are now in progress and which undoubtedly have taken place. Ought to state the opinion of the immutability of species and the creation by so many separate acts of will of the Creator[1].

[1] ⟨From the back of MS.⟩ Effect of climate on stationary island and on continent, but continent once island. Moreover repeated oscillations fresh diffusion when non-united, then isolation, when rising again immigration prevented, new habitats formed, new species, when united free immigration, hence uniform characters. Hence more forms ⟨on ?⟩ the island. Mountain summits. Why not true species. First let us recall in Part I, conditions of variation: change of conditions during several generations, and if frequently altered so much better [perhaps excess of food]. Secondly, continued selection [while in wild state]. Thirdly, isolation in all or nearly all,—as well to recall advantages of.

[In continent, if we look to terrestrial animal, long continued change might go on, which would only cause change in numerical number ⟨? proportions⟩: if continued long enough might ultimately affect all, though to most continents ⟨there is⟩ chance of immigration. Some few of whole body of species must be long affected and entire selection working same way. But here isolation absent, without barrier, cut off such ⟨illegible⟩. We can see advantage of isolation. But let us take case of island thrown up by volcanic agency at some distances, here we should have occasional visitants, only in few numbers and exposed to new conditions and ⟨illegible⟩ more important,—a quite new grouping of organic beings, which would open out new sources of subsistence, or ⟨would⟩ control ⟨?⟩ old ones. The number would be few, can old have the very best opportunity. ⟨The conquest of the indigenes by introduced organisms shows that the indigenes were not perfectly adapted, see *Origin*, Ed. i. p. 390.⟩ Moreover as the island continued changing,—continued slow changes, river, marshes, lakes, mountains &c. &c., new races as successively formed and a fresh occasional visitant.

If island formed continent, some species would emerge and immigrate. Everyone admits continents. We can see why Galapagos and C. Verde differ ⟨see *Origin*, Ed. i. p. 398⟩], depressed and raised. We can see from this repeated action and the time required for a continent, why many more forms than in New Zealand ⟨see *Origin*, Ed. i. p. 389 for a comparison between New Zealand and the Cape⟩ no mammals or other classes ⟨see however, *Origin*, Ed. i. p. 393 for the case of the frog⟩. We can at once see how it comes when there has been an old channel of migration,—Cordilleras; we can see why Indian Asiatic Flora,—[why species] having a wide range gives better chance of some arriving at new points and being selected, and adapted to new ends. I need hardly remark no necessity for change.

§ VII. ⟨AFFINITIES AND CLASSIFICATION.⟩

Looking now to the affinities of organisms, without relation to their distribution, and taking all fossil and recent, we see the degrees of relationship are of different degrees and arbitrary,—sub-genera,—genera,—sub-families, families, orders and classes and kingdoms. The kind of classification which everyone feels is most correct is called the natural system, but no can define this. If we say with Whewell ⟨that we have an⟩ undefined instinct of the importance of organs[1], we have no means in lower

Finally, as continent (most extinction ⟨?⟩ during formation of continent) is formed after repeated elevation and depression, and interchange of species we might foretell much extinction, and that the survivor would belong to same type, as the extinct, in same manner as different part of same continent, which were once separated by space as they are by time ⟨see *Origin*, Ed. i. pp. 339 and 349⟩.

As all mammals have descended from one stock, we ought to expect that every continent has been at some time connected, hence obliteration of present ranges. I do not mean that the fossil mammifers found in S. America are the lineal successors ⟨ancestors⟩ of the present forms of S. America: for it is highly improbable that more than one or two cases (who will say how many races after Plata bones) should be found. I believe this from numbers, who have lived,—mere ⟨?⟩ chance of fewness. Moreover in every case from very existence of genera and species only few at one time will leave progeny, under form of new species, to distant ages; and the more distant the ages the fewer the progenitors. An observation may be here appended, bad chance of preservation on rising island, the nurseries of new species, appeal to experience ⟨see *Origin*, Ed. i. p. 292⟩. This observation may be extended, that in all cases, subsiding land must be, in early stages, less favourable to formation of new species; but it will isolate them, and then if land recommences rising how favourable. As preoccupation is bar to diffusion to species, so would it be to a selected variety. But it would not be if that variety was better fitted to some not fully occupied station; so during elevation or the formation of new stations, is scene for new species. But during elevation not favourable to preservation of fossil (except in caverns ⟨?⟩); when subsidence highly favourable in early stages to preservation of fossils; when subsidence, less sediment. So that our strata, as general rule will be the tomb of old species (not undergoing any change) when rising land the nursery. But if there be vestige will generally be preserved to future ages, the new ones will not be entombed till fresh subsidence supervenes. In this long gap we shall have no record: so that wonderful if we should get transitional forms. I do not mean every stage, for we cannot expect that, as before shown, until geologists will be prepared to say that although under unnaturally favourable condition we can trace in future ages short-horn and Herefordshire ⟨see note 2, p. 26⟩.

[1] After "organs" is inserted, apparently as an afterthought:—"no, and instance metamorphosis, afterwards explicable."

animals of saying which is most important, and yet everyone feels that some one system alone deserves to be called natural. The true relationship of organisms is brought before one by considering relations of analogy, an otter-like animal amongst mammalia and an otter amongst marsupials. In such cases external resemblance and habit of life and *the final end of whole organization* very strong, yet no relation[1]. Naturalists cannot avoid these terms of relation and affinity though they use them metaphorically. If used in simple earnestness the natural system ought to be a genealogical ⟨one⟩; and our knowledge of the points which are most easily affected in transmission are those which we least value in considering the natural system, and practically when we find they do vary we regard them of less value[2]. In classifying varieties the same language is used and the same kind of division: here also (in pine-apple)[3] we talk of the natural classification, overlooking similarity of the fruits, because whole plant differs. The origin of sub-genera, genera, &c., &c., is not difficult on notion of genealogical succession, and accords with what we know of similar gradations of affinity in domesticated organisms. In the same region the organic beings are ⟨illegible⟩ related to each other and the external conditions in many physical respects are allied[4] and their differences of same kind, and therefore when a new species has been selected and has obtained a place in the economy of nature, we

[1] For analogical resemblances see *Origin*, Ed. i. p. 427, vi. p. 582.

[2] "Practically when naturalists are at work, they do not trouble themselves about the physiological value of the characters....If they find a character nearly uniform,...they use it as one of high value," *Origin*, Ed. i. p. 417, vi. p. 573.

[3] "We are cautioned...not to class two varieties of the pine-apple together, merely because their fruit, though the most important part, happens to be nearly identical," *Origin*, Ed. i. p. 423, vi. p. 579.

[4] The whole of this passage is obscure, but the text is quite clear, except for one illegible word.

may suppose that generally it will tend to extend its range during geographical changes, and thus, becoming isolated and exposed to new conditions, will slightly alter and its structure by selection become slightly remodified, thus we should get species of a sub-genus and genus,—as varieties of merino-sheep,—varieties of British and Indian cattle. Fresh species might go on forming and others become extinct and all might become extinct, and then we should have ⟨an⟩ extinct genus; a case formerly mentioned, of which numerous cases occur in Palæontology. But more often the same advantages which caused the new species to spread and become modified into several species would favour some of the species being preserved: and if two of the species, considerably different, each gave rise to group of new species, you would have two genera; the same thing will go on. We may look at case in other way, looking to future. According to mere chance every existing species may generate another, but if any species, A, in changing gets an advantage and that advantage (whatever it may be, intellect, &c., &c., or some particular structure or constitution) is inherited[1], A will be the progenitor of several genera or even families in the hard struggle of nature. A will go on beating out other forms, it might come that A would people earth,—we may now not have one descendant on our globe of the one or several original creations[2]. External conditions air, earth, water being same[3] on globe, and the communication not being perfect, organisms of widely different descent might become adapted to

[1] ⟨The exact position of the following passage is uncertain:⟩ "just as it is not likely every present breed of fancy birds and cattle will propagate, only some of the best."

[2] This suggests that the author was not far from the principle of divergence on which he afterwards laid so much stress. See *Origin*, Ed. i. p. 111, vi. p. 134, also *Life and Letters*, i. p. 84.

[3] That is to say, the same conditions occurring in different parts of the globe.

the same end and then we should have cases of analogy[1], [they might even tend to become numerically representative]. From this often happening each of the great divisions of nature would have their representative eminently adapted to earth, to ⟨air⟩[2], to water, and to these in ⟨illegible⟩ and then these great divisions would show numerical relations in their classification.

§ VIII. UNITY [OR SIMILARITY] OF TYPE IN THE GREAT CLASSES.

Nothing more wonderful in Nat. Hist. than looking at the vast number of organisms, recent and fossil, exposed to the most diverse conditions, living in the most distant climes, and at immensely remote periods, fitted to wholely different ends, yet to find large groups united by a similar type of structure. When we for instance see bat, horse, porpoise-fin, hand, all built on same structure[3], having bones[4] with same name, we see there is some deep bond of union between them[5], to illustrate this is the foundation and objects ⟨?⟩ ⟨of⟩ what is called the Natural System; and which is foundation of distinction ⟨?⟩ of true and adaptive characters[6]. Now this wonderful fact of hand, hoof, wing, paddle and claw being the same, is at once explicable on the principle of some parent-forms, which might either be ⟨illegible⟩ or walking animals, becoming through infinite number of small

[1] The position of the following is uncertain, "greyhound and racehorse have an analogy to each other." The same comparison occurs in the *Origin*, Ed. i. p. 427, vi. p. 583.

[2] *Air* is evidently intended; in the MS. *water* is written twice.

[3] Written between the lines occurs:—"extend to birds and other classes."

[4] Written between the lines occurs:—"many bones merely represented."

[5] In the *Origin*, Ed. i. p. 434, vi. p. 595, the term *morphology* is taken as including *unity of type*. The paddle of the porpoise and the wing of the bat are there used as instances of morphological resemblance.

[6] The sentence is difficult to decipher.

selections adapted to various conditions. We know that proportion, size, shape of bones and their accompanying soft parts vary, and hence constant selection would alter, to almost any purpose ⟨?⟩ the framework of an organism, but yet would leave a general, even closest similarity in it.

[We know the number of similar parts, as vertebræ and ribs can vary, hence this also we might expect.] Also ⟨if⟩ the changes carried on to a certain point, doubtless type will be lost, and this is case with Plesiosaurus[1]. The unity of type in past and present ages of certain great divisions thus undoubtedly receives the simplest explanation.

There is another class of allied and almost identical facts, admitted by the soberest physiologists, [from the study of a certain set of organs in a group of organisms] and refers ⟨? referring⟩ to a unity of type of different organs in the same individual, denominated the science of "Morphology." The ⟨? this⟩ discovered by beautiful and regular series, and in the case of plants from monstrous changes, that certain organs in an individual are other organs metamorphosed. Thus every botanist considers petals, nectaries, stamens, pistils, germen as metamorphosed leaf. They thus explain, in the most lucid manner, the position and number of all parts of the flower, and the curious conversion under cultivation of one part into another. The complicated double set of jaws and palpi of crustaceans[2], and all insects are considered as metamorphosed ⟨limbs⟩ and to see the series is to admit this phraseology. The skulls of the vertebrates are undoubtedly composed of three metamorphosed vertebræ; thus we can understand the strange form of

[1] In the *Origin*, Ed. i. p. 436, vi. p. 598, the author speaks of the "general pattern" being obscured in the paddles of "extinct gigantic sea-lizards."

[2] See *Origin*, Ed. i. p. 437, vi. p. 599.

the separate bones which compose the casket holding man's brain. These[1] facts differ but slightly from those of last section, if with wing, paddle, hand and hoof, some common structure was yet visible, or could be made out by a series of occasional monstrous conversions, and if traces could be discovered of ⟨the⟩ whole having once existed as walking or swimming instruments, these organs would be said to be metamorphosed, as it is they are only said to exhibit a common type.

This distinction is not drawn by physiologists, and is only implied by some by their general manner of writing. These facts, though affecting every organic being on the face of the globe, which has existed, or does exist, can only be viewed by the Creationist as ultimate and inexplicable facts. But this unity of type through the individuals of a group, and this metamorphosis of the same organ into other organs, adapted to diverse use, necessarily follows on the theory of descent[2]. For let us take case of Vertebrata, which if[3] they descended from one parent and by this theory all the Vertebrata have been altered by slow degrees, such as we see in domestic animals. We know that proportions alter, and even that occasionally numbers of vertebræ alter, that parts become soldered, that parts are lost, as tail and toes, but we know ⟨that ?⟩ here we can see that possibly a walking organ might ⟨?⟩ be converted into swimming or into a gliding organ and so on to a flying organ. But such gradual changes would not alter the unity of type in their descendants, as parts lost and soldered and vertebræ.

[1] The following passage seems to have been meant to precede the sentence beginning "These facts":—"It is evident, that when in each individual species, organs are metamorph. a unity of type extends."

[2] This is, I believe, the first place in which the author uses the words "theory of descent."

[3] The sentence should probably run, "Let us take the case of the vertebrata: if we assume them to be descended from one parent, then by this theory they have been altered &c."

But we can see that if this carried to extreme, unity lost,—Plesiosaurus. Here we have seen the same organ is formed ⟨?⟩ ⟨for⟩ different purposes ⟨ten words illegible⟩: and if, in several orders of vertebrata, we could trace origin ⟨of⟩ spinous processes and monstrosities &c. we should say, instead of there existing a unity of type, morphology[1], as we do when we trace the head as being the vertebræ metamorphosed. Be it observed that Naturalists, as they use terms of affinity without attaching real meaning, here also they are obliged to use metamorphosis, without meaning that any parent of crustacean was really an animal with as many legs as crustacean has jaws. The theory of descent at once explains these wonderful facts.

Now few of the physiologists who use this language really suppose that the parent of insect with the metamorphosed jaw, was an insect with [more] so many legs, or that the parent of flowering plants, originally had no stamens, or pistils or petals, but some other means of propagation,—and so in other cases. Now according to our theory during the infinite number of changes, we might expect that an organ used for a purpose might be used for a different one by his descendant, as must have been the case by our theory with the bat, porpoise, horse, &c., which are descended from one parent. And if it so chanced that traces of the former use and structure of the part should be retained, which is manifestly possible if not probable, then we should have the organs, on which morphology is founded and which instead of being metaphorical becomes plain and ⟨and instead of being⟩ utterly unintelligible becomes simple matter of fact[2].

1 That is "we should call it a morphological fact."

2 In the *Origin*, Ed. i. p. 438, vi. p. 602, the author, referring to the expressions used by naturalists in regard to morphology and metamorphosis, says "On my view these terms may be used literally."

⟨*Embryology.*⟩ This general unity of type in great groups of organisms (including of course these morphological cases) displays itself in a most striking manner in the stages through which the fœtus passes[1]. In early stage, the wing of bat, hoof, hand, paddle are not to be distinguished. At a still earlier ⟨stage⟩ there is no difference between fish, bird, &c. &c. and mammal. It is not that they cannot be distinguished, but the arteries[2] ⟨illegible⟩. It is not true that one passes through the form of a lower group, though no doubt fish more nearly related to fœtal state[3].

This similarity at the earliest stage is remarkably shown in the course of the arteries which become greatly altered, as fœtus advances in life and assumes the widely different course and number which characterize full-grown fish and mammals. How wonderful that in egg, in water or air, or in womb of mother, artery[4] should run in same course.

Light can be thrown on this by our theory. The structure of each organism is chiefly adapted to the sustension of its life, when full-grown, when it has to feed itself and propagate[5]. The structure of a kitten is quite in secondary degree adapted to its habits, whilst fed by its mother's milk and prey. Hence variation in the structure of the full-grown species will *chiefly* determine the preservation of a

[1] See *Origin*, Ed. i. p. 439, vi. p. 605.

[2] In the *Origin*, Ed. i. p. 440, vi. p. 606, the author argues that the "loop-like course of the arteries" in the vertebrate embryo has no direct relation to the conditions of existence.

[3] The following passages are written across the page:—"They pass through the same phases, but some, generally called the higher groups, are further metamorphosed.

? Degradation and complication ? no tendency to perfection.

? Justly argued against Lamarck ?"

[4] An almost identical passage occurs in the *Origin*, Ed. i. p. 440, vi. p. 606.

[5] The following: "Deaths of brothers ⟨when⟩ old by same peculiar disease" which is written between the lines seems to have been a memorandum which is expanded a few lines lower. I believe the case of the brothers came from Dr R. W. Darwin.

species now become ill-suited to its habitat, or rather with a better place opened to it in the economy of Nature. It would not matter to the full-grown cat whether in its young state it was more or less eminently feline, so that it become so when full-grown. No doubt most variation, (not depending on habits of life of individual) depends on early change[1] and we must suspect that at whatever time of life the alteration of fœtus is effected, it tends to appear at same period. When we ⟨see⟩ a tendency to particular disease in old age transmitted by the male, we know some effect is produced during conception, on the simple cell of ovule, which will not produce its effect till half a century afterwards and that effect is not visible[2]. So we see in grey-hound, bull-dog, in race-horse and cart-horse, which have been selected for their form in full-life, there is much less (?) difference in the few first days after birth[3], than when full-grown: so in cattle, we see it clearly in cases of cattle, which differ obviously in shape and length of horns. If man were during 10,000 years to be able to select, far more diverse animals from horse or cow, I should expect there would be far less differences in the very young and fœtal state: and this, I think, throws light on above marvellous fact. In larvæ, which have long life selection, perhaps, does much,—in the pupa not so much[4]. There is no

[1] See the discussion to this effect in the *Origin*, Ed. i. pp. 443–4, vi. p. 610. The author there makes the distinction between a cause affecting the germ-cell and the reaction occurring at a late period of life.

[2] Possibly the sentence was meant to end "is not visible till then."

[3] See *Origin*, Ed. i. pp. 444–5, vi. p. 611. The query appended to *much less* is justified, since measurement was necessary to prove that the grey-hound and bulldog puppies had not nearly acquired "their full amount of proportional difference."

[4] ⟨The following discussion, from the back of the page, is in large measure the same as the text.⟩ I think light can be thrown on these facts. From the following peculiarities being hereditary, [we know that some change in the germinal vesicle is effected, which will only betray itself years after] diseases —man, goitre, gout, baldness, fatness, size, [longevity ⟨illegible⟩ time of

object gained in varying form &c. of fœtus (beyond certain adaptations to mother's womb) and therefore selection will not further act on it, than in giving to its changing tissues a tendency to certain parts afterwards to assume certain forms.

Thus there is no power to change the course of

reproduction, shape of horns, case of old brothers dying of same disease]. And we know that the germinal vesicle must have been affected, though no effect is apparent or can be apparent till years afterwards,—no more apparent than when these peculiarities appear by the exposure of the full-grown individual. ⟨That is, "the young individual is as apparently free from the hereditary changes which will appear later, as the young is actually free from the changes produced by exposure to certain conditions in adult life."⟩ So that when we see a variety in cattle, even if the variety be due to act of reproduction, we cannot feel sure at what period this change became apparent. It may have been effected during early age of free life ⟨or⟩ fœtal existence, as monsters show. From arguments before used, and crossing, we may generally suspect in germ; but I repeat it does not follow, that the change should be apparent till life fully developed; any more than fatness depending on heredity should be apparent during early childhood, still less during fœtal existence. In case of horns of cattle, which when inherited must depend on germinal vesicle, obviously no effect till cattle full-grown. Practically it would appear that the [hereditary] peculiarities characterising our domestic races, therefore resulting from vesicle, do not appear with their full characters in very early states; thus though two breeds of cows have calves different, they are not so different,—grey-hound and bull-dog. And this is what is ⟨to⟩ be expected, for man is indifferent to characters of young animals and hence would select those full-grown animals which possessed the desirable characteristics. So that from mere chance we might expect that some of the characters would be such only as became fully apparent in mature life. Furthermore we may suspect it to be a law, that at whatever time a new character appears, whether from vesicle, or effects of external conditions, it would appear at corresponding time ⟨see *Origin*, Ed. i. p. 444⟩. Thus diseases appearing in old age produce children with d°.,—early maturity,—longevity,—old men, brothers, of same disease—young children of d°. I said men do not select for quality of young,—calf with big bullocks. Silk-worms, peculiarities which, appear in caterpillar state or cocoon state, are transmitted to corresponding states. The effect of this would be that if some peculiarity was born in a young animal, but never exercised, it might be inherited in young animal; but if exercised that part of structure would be increased and would be inherited in corresponding time of life after such training.

I have said that man selects in full-life, so would it be in Nature. In struggle of existence, it matters nothing to a feline animal, whether kitten eminently feline, as long as it sucks. Therefore natural selection would act equally well on character which was fully ⟨developed⟩ only in full age. Selection could tend to alter no character in fœtus, (except relation to mother) it would alter less in young state (putting on one side larva condition) but alter every part in full-grown condition. Look to a fœtus and its parent, and again after ages fœtus and its ⟨i.e. the above mentioned parents⟩ descendant; the parent more variable ⟨?⟩ than fœtus, which explains all.

the arteries, as long as they nourish the fœtus; it is the selection of slight changes which supervene at any time during ⟨illegible⟩ of life.

The less differences of fœtus,—this has obvious meaning on this view: otherwise how strange that a [monkey] horse, a man, a bat should at one time of life have arteries, running in a manner, which is only intelligibly useful in a fish! The natural system being on theory genealogical, we can at once see, why fœtus, retaining traces of the ancestral form, is of the highest value in classification.

§ IX. ⟨Abortive organs.⟩

There is another grand class of facts relating to what are called abortive organs. These consist of organs which the same reasoning power that shows us how beautifully these organs in some cases are adapted to certain end, declares in other cases are absolutely useless. Thus teeth in Rhinoceros[1], whale, narwhal,—bone on tibia, muscles which do not move,—little bone of wing of Apteryx,—bone representing extremities in some snake,—little wings within ⟨?⟩ soldered cover of beetles,—men and bulls, mammæ: filaments without anthers in plants, mere scales representing petals in others, in feather-hyacinth whole flower. Almost infinitely numerous. No one can reflect on these without astonishment, can anything be clearer than that wings are to fly and teeth ⟨to bite⟩, and yet we find these organs perfect in every detail in situations where they cannot possibly be of their normal use[2].

The term abortive organ has been thus applied

[1] Some of these examples occur in *Origin*, Ed. i. pp. 450–51, vi. pp. 619–20.

[2] The two following sentences are written, one down the margin, the other across the page. "Abortive organs eminently useful in classification. Embryonic state of organs. Rudiments of organs."

to above structure (as *invariable* as all other parts[1]) from their absolute similarity to monstrous cases, where from *accident*, certain organs are not developed; as infant without arms or fingers with mere stump representing them: teeth represented by mere points of ossification: headless children with mere button,—viscera represented by small amorphous masses, &c.,—the tail by mere stump,—a solid horn by minute hanging one[2]. There is a tendency in all these cases, when life is preserved, for such structures to become hereditary. We see it in tailless dogs and cats. In plants we see this strikingly,—in Thyme, in *Linum flavum*,—stamen in *Geranium pyrenaicum*[3]. Nectaries abort into petals in Columbine ⟨*Aquilegia*⟩, produced from some accident and then become hereditary, in some cases only when propagated by buds, in other cases by seed. These cases have been produced suddenly by accident in early growth, but it is part of law of growth that when any organ is not used it tends to diminish (duck's wing[4]?) muscles of dog's ears, ⟨and of⟩ rabbits, muscles wither, arteries grow up. When eye born defective, optic nerve (Tuco Tuco) is atrophied. As every part whether useful or not (diseases, double flowers) tends to be transmitted to offspring, the origin of abortive organs whether produced at the birth or slowly acquired is easily understood in domestic races of organisms: [a struggle between the atrophy and hereditariness. Abortive organs in domestic races.] There will always be a struggle between atrophy of an organ rendered useless, and

[1] I imagine the meaning to be that abortive organs are specific characters in contrast to monstrosities.

[2] Minute hanging horns are mentioned in the *Origin*, Ed. i. p. 454, vi. p. 625, as occurring in hornless breeds of cattle.

[3] *Linum flavum* is dimorphic: thyme gynodiæcious. It is not clear what point is referred to under *Geranium pyrenaicum*.

[4] The author's work on duck's wings &c. is in *Var. under Dom.*, Ed. 2, i. p. 299.

hereditariness[1]. Because we can understand the origin of abortive organs in certain cases, it would be wrong to conclude absolutely that all must have had same origin, but the strongest analogy is in favour of it. And we can by our theory, for during infinite changes some organ, we might have anticipated, would have become useless. ⟨We can⟩ readily explain the fact, so astounding on any other view, namely that organs possibly useless have been formed often with the same exquisite care as when of vital importance.

Our theory, I may remark would permit an organ ⟨to⟩ become abortive with respect to its primary use, to be turned to any other purpose, (as the buds in a cauliflower) thus we can see no difficulty in bones of male marsupials being used as fulcrum of muscles, or style of marygold[2],—indeed in one point of view, the heads of [vertebrated] animal may be said to be abortive vertebræ turned into other use: legs of some crustacea abortive jaws, &c., &c. De Candolle's analogy of table covered with dishes[3].

⟨The following passage was possibly intended to be inserted here.⟩ Degradation and complication see Lamarck: no tendency to perfection: if room, [even] high organism would have greater power in beating lower one, thought ⟨?⟩ to be selected for a degraded end.

[1] The words *vis medicatrix* are inserted after "useless," apparently as a memorandum.

[2] In the male florets of certain Compositæ the style functions merely as a piston for forcing out the pollen.

[3] ⟨On the back of the page is the following.⟩ If abortive organs are a trace preserved by hereditary tendency, of organ in ancestor of use, we can at once see why important in natural classification, also why more plain in young animal because, as in last section, the selection has altered the old animal most. I repeat, these wondrous facts, of parts created for no use in past and present time, all can by my theory receive simple explanation; or they receive none and we must be content with some such empty metaphor, as that of De Candolle, who compares creation to a well covered table, and says abortive organs may be compared to the dishes (some should be empty) placed symmetrically!

§ x. Recapitulation and conclusion.

Let us recapitulate the whole ⟨?⟩ ⟨of⟩ these latter sections by taking case of the three species of Rhinoceros, which inhabit Java, Sumatra, and mainland of Malacca or India. We find these three close neighbours, occupants of distinct but neighbouring districts, as a group having a different aspect from the Rhinoceros of Africa, though some of these latter inhabit very similar countries, but others most diverse stations. We find them intimately related [scarcely ⟨?⟩ differences more than some breeds of cattle] in structure to the Rhinoceros, which for immense periods have inhabited this one, out of three main zoological divisions of the world. Yet some of these ancient animals were fitted to very different stations: we find all three ⟨illegible⟩ of the generic character of the Rhinoceros, which form a [piece of net][1] set of links in the broken chain representing the Pachydermata, as the chain likewise forms a portion in other and longer chains. We see this wonderfully in dissecting the coarse leg of all three and finding nearly the same bones as in bat's wings or man's hand, but we see the clear mark in solid tibia of the fusion into it of the fibula. In all three we find their heads composed of three altered vertebræ, short neck, same bones as giraffe. In the upper jaws of all three we find small teeth like rabbit's. In dissecting them in fœtal state we find at a not very early stage their form exactly alike the most different animals, and even with arteries running as in a fish: and this similarity holds when the young one is produced in womb, pond, egg or spawn. Now these three undoubted species scarcely differ more than breeds of cattle,

[1] The author doubtless meant that the complex relationships between organisms can be roughly represented by a net in which the knots stand for species.

are probably subject to many the same contagious diseases; if domesticated these forms would vary, and they might possibly breed together, and fuse into something[1] different ⟨from⟩ their aboriginal forms; might be selected to serve different ends.

Now the Creationist believes these three Rhinoceroses were created[2] with their deceptive appearance of true, not ⟨illegible⟩ relationship; as well can I believe the planets revolve in their present courses not from one law of gravity but from distinct volition of Creator.

If real species, sterile one with another, differently adapted, now inhabiting different countries, with different structures and instincts, are admitted to have common descent, we can only legitimately stop where our facts stop. Look how far in some case a chain of species will lead us. ⟨This probably refers to the Crustacea, where the two ends of the series have "hardly a character in common." *Origin*, Ed. i. p. 419.⟩ May we not jump (considering how much extermination, and how imperfect geological records) from one sub-genus to another sub-genus. Can genera restrain us; many of the same arguments, which made us give up species, inexorably demand genera and families and orders to fall, and classes tottering. We ought to stop only when clear unity of type, independent of use and adaptation, ceases.

Be it remembered no naturalist pretends to give test from external characters of species; in many genera the distinction is quite arbitrary[3]. But there remains one other way of comparing species

[1] Between the lines occurs :—"one ⟨?⟩ form be lost."

[2] The original sentence is here broken up by the insertion of :—"out of the dust of Java, Sumatra, these ⟨?⟩ allied to past and present age and ⟨illegible⟩, with the stamp of inutility in some of their organs and conversion in others."

[3] Between the lines occur the words :—"Species vary according to same general laws as varieties ; they cross according to same laws."

with races; it is to compare the effects of crossing them. Would it not be wonderful, if the union of two organisms, produced by two separate acts of Creation, blended their characters together when crossed according to the same rules, as two races which have undoubtedly descended from same parent stock; yet this can be shown to be the case. For sterility, though a usual ⟨?⟩, is not an invariable concomitant, it varies much in degree and has been shown to be probably dependent on causes closely analogous with those which make domesticated organisms sterile. Independent of sterility there is no difference between mongrels and hybrids, as can be shown in a long series of facts. It is strikingly seen in cases of instincts, when the minds of the two species or races become blended together[1]. In both cases if the half-breed be crossed with either parent for a few generations, all traces of the one parent form is lost (as Kölreuter in two tobacco species almost sterile together), so that the Creationist in the case of a species, must believe that one act of creation is absorbed into another!

Conclusion.

Such are my reasons for believing that specific forms are not immutable. The affinity of different groups, the unity of types of structure, the representative forms through which fœtus passes, the metamorphosis of organs, the abortion of others cease to be metaphorical expressions and become intelligible facts. We no longer look ⟨an⟩ on animal as a savage does at a ship[2], or other great work of art, as a thing wholly beyond comprehension, but we

[1] "A cross with a bull-dog has affected for many generations the courage and obstinacy of greyhounds," *Origin*, Ed. i. p. 214, vi. p. 327.

[2] The simile of the savage and the ship occurs in the *Origin*, Ed. i. p. 485, vi. p. 665.

Conclusion

Such are my reasons for believing that specific forms are not immutable. The affinity of different groups, the unity of types of structure, the representative forms through which foetus passes, the metamorphosis of organs, the abortion of others cease to be metaphorical expressions and become intelligible facts. We no longer look at an animal as a savage does at a ship or other great work of art, as a thing wholly beyond comprehension, but we feel far more interest in examining it.

Facsimile of the original manuscript of the paragraph on p. 50.

feel far more interest in examining it. How interesting is every instinct, when we speculate on their origin as an hereditary or congenital habit or produced by the selection of individuals differing slightly from their parents. We must look at every complicated mechanism and instinct, as the summary of a long history, ⟨as the summing up⟩ of[1] useful contrivances, much like a work of art. How interesting does the distribution of all animals become, as throwing light on ancient geography. [We see some seas bridged over.] Geology loses in its glory from the imperfection of its archives[2], but how does it gain in the immensity of the periods of its formations and of the gaps separating these formations. There is much grandeur in looking at the existing animals either as the lineal descendants of the forms buried under thousand feet of matter, or as the coheirs of some still more ancient ancestor. It accords with what we know of the law impressed on matter by the Creator, that the creation and extinction of forms, like the birth and death of individuals should be the effect of secondary [laws] means[3]. It is derogatory that the Creator of countless systems of worlds should have created each of the myriads of creeping parasites and [slimy] worms which have swarmed each day of life on land and water ⟨on⟩ [this] one globe. We cease being astonished, however much we may deplore, that a group of animals should have been directly created to lay their eggs in bowels and flesh of other,—that some organisms should delight in cruelty,—that animals should be led away by false instincts,—that annually there should be an

[1] In the *Origin*, Ed. i. p. 486, vi. p. 665, the author speaks of the "summing up of many contrivances": I have therefore introduced the above words which make the passage clearer. In the *Origin* the comparison is with "a great mechanical invention,"—not with a work of art.

[2] See a similar passage in the *Origin*, Ed. i. p. 487, vi. p. 667.

[3] See the *Origin*, Ed. i. p. 488, vi. p. 668.

incalculable waste of eggs and pollen. From death, famine, rapine, and the concealed war of nature we can see that the highest good, which we can conceive, the creation of the higher animals has directly come. Doubtless it at first transcends our humble powers, to conceive laws capable of creating individual organisms, each characterised by the most exquisite workmanship and widely-extended adaptations. It accords better with [our modesty] the lowness of our faculties to suppose each must require the fiat of a creator, but in the same proportion the existence of such laws should exalt our notion of the power of the omniscient Creator[1]. There is a simple grandeur in the view of life with its powers of growth, assimilation and reproduction, being originally breathed into matter under one or a few forms, and that whilst this our planet has gone circling on according to fixed laws, and land and water, in a cycle of change, have gone on replacing each other, that from so simple an origin, through the process of gradual selection of infinitesimal changes, endless forms most beautiful and most wonderful have been evolved[2].

[1] The following discussion, together with some memoranda are on the last page of the MS. "The supposed creative spirit does not create either number or kind which ⟨are⟩ from analogy adapted to site (viz. New Zealand): it does not keep them all permanently adapted to any country,—it works on spots or areas of creation,—it is not persistent for great periods,—it creates forms of same groups in same regions, with no physical similarity,—it creates, on islands or mountain summits, species allied to the neighbouring ones, and not allied to alpine nature as shown in other mountain summits —even different on different island of similarly constituted archipelago, not created on two points: never mammifers created on small isolated island; nor number of organisms adapted to locality: its power seems influenced or related to the range of other species wholly distinct of the same genus,—it does not equally effect, in amount of difference, all the groups of the same class."

[2] This passage is the ancestor of the concluding words in the first edition of the *Origin of Species* which have remained substantially unchanged throughout subsequent editions, "There is grandeur in this view of life, with its several powers, having been originally breathed into a few forms or into one; and that whilst this planet has gone cycling on according to the fixed law of gravity, from so simple a beginning endless forms most

N.B.—There ought somewhere to be a discussion from Lyell to show that external conditions do vary, or a note to Lyell's works (work ?).

Besides other difficulties in ii. Part, non-acclimatisation of plants. Difficulty when asked *how* did white and negro become altered from common intermediate stock: no facts. We do NOT know that species are immutable, on the contrary. What arguments against this theory, except our not perceiving every step, like the erosion of valleys[1].

beautiful and most wonderful have been, and are being, evolved." In the 2nd edition "by the Creator" is introduced after "originally breathed."

[1] Compare the *Origin*, Ed. i. p. 481, vi. p. 659, "The difficulty is the same as that felt by so many geologists, when Lyell first insisted that long lines of inland cliffs had been formed, and great valleys excavated, by the slow action of the coast-waves."

THE ESSAY OF 1844

PART I

CHAPTER I

ON THE VARIATION OF ORGANIC BEINGS UNDER DOMESTICATION; AND ON THE PRINCIPLES OF SELECTION

THE most favourable conditions for variation seem to be when organic beings are bred for many generations under domestication[1]: one may infer this from the simple fact of the vast number of races and breeds of almost every plant and animal, which has long been domesticated. Under certain conditions organic beings even during their individual lives become slightly altered from their usual form, size, or other characters: and many of the peculiarities thus acquired are transmitted to their offspring. Thus in animals, the size and vigour of body, fatness, period of maturity, habits of body or consensual movements, habits of mind and temper, are modified or acquired during the life of the individual[2], and become inherited. There is reason to believe that when long exercise has given to certain muscles great development, or disuse has lessened them, that such development is also in-

[1] The cumulative effect of domestication is insisted on in the *Origin*, see *e.g. Origin*, Ed. i. p. 7, vi. p. 8.

[2] This type of variation passes into what he describes as the direct effect of conditions. Since they are due to causes acting during the adult life of the organism they might be called individual variations, but he uses this term for congenital variations, *e.g.* the differences discoverable in plants raised from seeds of the same pod (*Origin*, Ed. i. p. 45, vi. p. 53).

herited. Food and climate will occasionally produce changes in the colour and texture of the external coverings of animals; and certain unknown conditions affect the horns of cattle in parts of Abyssinia; but whether these peculiarities, thus acquired during individual lives, have been inherited, I do not know. It appears certain that malconformation and lameness in horses, produced by too much work on hard roads,—that affections of the eyes in this animal probably caused by bad ventilation,—that tendencies towards many diseases in man, such as gout, caused by the course of life and ultimately producing changes of structure, and that many other diseases produced by unknown agencies, such as goitre, and the idiotcy resulting from it, all become hereditary.

It is very doubtful whether the flowers and leaf-buds, annually produced from the same bulb, root, or tree, can properly be considered as parts of the same individual, though in some respects they certainly seem to be so. If they are parts of an individual, plants also are subject to considerable changes during their *individual* lives. Most florist-flowers if neglected degenerate, that is, they lose some of their characters; so common is this, that trueness is often stated, as greatly enhancing the value of a variety[1]: tulips break their colours only after some years' culture; some plants become double and others single, by neglect or care: these characters can be transmitted by cuttings or grafts, and in some cases by true or seminal propagation. Occasionally a single bud on a plant assumes at once a new and widely different character: thus it is certain that nectarines have been produced on

[1] ⟨It is not clear where the following note is meant to come⟩: Case of Orchis,—most remarkable as not long cultivated by seminal propagation. Case of varieties which soon acquire, like *Ægilops* and Carrot (and Maize) *a certain general character* and then go on varying.

peach trees and moss roses on provence roses; white currants on red currant bushes; flowers of a different colour from that of the stock, in Chrysanthemums, Dahlias, sweet-williams, Azaleas, &c., &c.; variegated leaf-buds on many trees, and other similar cases. These new characters appearing in single buds, can, like those lesser changes affecting the whole plant, be multiplied not only by cuttings and such means, but often likewise by true seminal generation.

The changes thus appearing during the lives of individual animals and plants are extremely rare compared with those which are congenital or which appear soon after birth. Slight differences thus arising are infinitely numerous: the proportions and form of every part of the frame, inside and outside, appear to vary in very slight degrees: anatomists dispute what is the "beau ideal" of the bones, the liver and kidneys, like painters do of the proportions of the face: the proverbial expression that no two animals or plants are born absolutely alike, is much truer when applied to those under domestication, than to those in a state of nature[1]. Besides these slight differences, single individuals are occasionally born considerably unlike in certain parts or in their whole structure to their parents: these are called by horticulturalists and breeders "sports"; and are not uncommon except when very strongly marked. Such sports are known in some cases to have been parents of some of our domestic races; and such probably have been the parents of many other races, especially of those which in some senses may be called hereditary monsters; for instance where there is an additional limb, or where all the limbs are stunted (as in the Ancon sheep), or where a part is wanting, as in rumpless fowls and tailless

[1] Here, as in the MS. of 1842, the author is inclined to minimise the variation occurring in nature.

dogs or cats[1]. The effects of external conditions on the size, colour and form, which can rarely and obscurely be detected during one individual life, become apparent after several generations: the slight differences, often hardly describable, which characterize the stock of different countries, and even of districts in the same country, seem to be due to such continued action.

On the hereditary tendency.

A volume might be filled with facts showing what a strong tendency there is to inheritance, in almost every case of the most trifling, as well as of the most remarkable congenital peculiarities[2]. The term congenital peculiarity, I may remark, is a loose expression and can only mean a peculiarity apparent when the part affected is nearly or fully developed: in the Second Part, I shall have to discuss at what period of the embryonic life connatal peculiarities probably first appear; and I shall then be able to show from some evidence, that at whatever period of life a new peculiarity first appears, it tends hereditarily to appear at a corresponding period[3]. Numerous though slight changes, slowly supervening in animals during mature life (often, though by no means always, taking the form of disease), are, as stated in the first paragraphs, very often hereditary. In plants, again, the buds which assume a different character from their stock likewise tend to transmit their new peculiarities. There is not sufficient reason to believe that either mutilations[4] or changes of form produced by

[1] This is more strongly stated than in the *Origin*, Ed. i. p. 30.

[2] See *Origin*, Ed. i. p. 13.

[3] *Origin*, Ed. i. p. 86, vi. p. 105.

[4] It is interesting to find that though the author, like his contemporaries, believed in the inheritance of acquired characters, he excluded the case of mutilation.

mechanical pressure, even if continued for hundreds of generations, or that any changes of structure quickly produced by disease, are inherited; it would appear as if the tissue of the part affected must slowly and freely grow into the new form, in order to be inheritable. There is a very great difference in the hereditary tendency of different peculiarities, and of the same peculiarity, in different individuals and species; thus twenty thousand seeds of the weeping ash have been sown and not one come up true;—out of seventeen seeds of the weeping yew, nearly all came up true. The ill-formed and almost monstrous "Niata" cattle of S. America and Ancon sheep, both when bred together and when crossed with other breeds, seem to transmit their peculiarities to their offspring as truly as the ordinary breeds. I can throw no light on these differences in the power of hereditary transmission. Breeders believe, and apparently with good cause, that a peculiarity generally becomes more firmly implanted after having passed through several generations; that is if one offspring out of twenty inherits a peculiarity from its parents, then its descendants will tend to transmit this peculiarity to a larger proportion than one in twenty; and so on in succeeding generations. I have said nothing about mental peculiarities being inheritable for I reserve this subject for a separate chapter.

Causes of Variation.

Attention must here be drawn to an important distinction in the first origin or appearance of varieties: when we see an animal highly kept producing offspring with an hereditary tendency to early maturity and fatness; when we see the wild-duck and Australian dog always becoming, when bred for one or a few generations in confinement, mottled in their colours; when we see people living in certain districts or circumstances becoming subject to an hereditary taint to certain organic diseases, as consumption or plica polonica,—we naturally attribute such changes to the direct effect of known or unknown agencies acting for one or more generations on the parents. It is probable that a multitude of peculiarities may be thus directly caused by unknown external agencies. But in breeds, characterized by an extra limb or claw, as in certain fowls and dogs; by an extra joint in the vertebræ; by the loss of a part, as the tail; by the substitution of a tuft of feathers for a comb in certain poultry; and in a multitude of other cases, we can hardly attribute these peculiarities directly to external influences, but indirectly to the laws of embryonic growth and of reproduction. When we see a multitude of varieties (as has often been the case, where a cross has been carefully guarded against) produced from seeds matured in the very same capsule[1], with the male and female principle nourished from the same roots and necessarily exposed to the same external influences; we cannot believe that the endless slight differences between seedling varieties thus produced, can be the effect of any corresponding difference in their exposure. We are led (as Müller has remarked) to the same conclusion, when we see in the same litter, produced by the same act of conception, animals considerably different.

As variation to the degree here alluded to has been observed only in organic beings under domestication, and in plants amongst those most highly and long cultivated, we must attribute, in such cases, the varieties (although the difference between each variety cannot possibly be attributed to any corresponding difference of exposure in the parents) to the indirect effects of domestication on the action of

[1] This corresponds to *Origin*, Ed. i. p. 10, vi. p. 9.

the reproductive system[1]. It would appear as if the reproductive powers failed in their ordinary function of producing new organic beings closely like their parents; and as if the entire organization of the embryo, under domestication, became in a slight degree plastic[2]. We shall hereafter have occasion to show, that in organic beings, a considerable change from the natural conditions of life, affects, independently of their general state of health, in another and remarkable manner the reproductive system. I may add, judging from the vast number of new varieties of plants which have been produced in the same districts and under nearly the same routine of culture, that probably the indirect effects of domestication in making the organization plastic, is a much more efficient source of variation than any direct effect which external causes may have on the colour, texture, or form of each part. In the few instances in which, as in the Dahlia[3], the course of variation has been recorded, it appears that domestication produces little effect for several generations in rendering the organization plastic; but afterwards, as if by an accumulated effect, the original character of the species suddenly gives way or breaks.

On Selection.

We have hitherto only referred to the first appearance in individuals of new peculiarities; but to make a race or breed, something more is generally[4] requisite than such peculiarities (except

1 *Origin*, Ed. i. p. 8, vi. p. 10.

2 For *plasticity* see *Origin*, Ed. i. pp. 12, 132.

3 *Var. under Dom.*, Ed. ii. I. p. 393.

4 Selection is here used in the sense of isolation, rather than as implying the summation of small differences. Professor Henslow in his *Heredity of Acquired Characters in Plants*, 1908, p. 2, quotes from Darwin's *Var. under Dom.*, Ed. i. II. p. 271, a passage in which the author, speaking of the direct action of conditions, says:—"A new sub-variety would thus be produced without the aid of selection." Darwin certainly

in the case of the peculiarities being the direct effect of constantly surrounding conditions) should be inheritable,—namely the principle of selection, implying separation. Even in the rare instances of sports, with the hereditary tendency very strongly implanted, crossing must be prevented with other breeds, or if not prevented the best characterized of the half-bred offspring must be carefully selected. Where the external conditions are constantly tending to give some character, a race possessing this character will be formed with far greater ease by selecting and breeding together the individuals most affected. In the case of the endless slight variations produced by the indirect effects of domestication on the action of the reproductive system, selection is indispensable to form races; and when carefully applied, wonderfully numerous and diverse races can be formed. Selection, though so simple in theory, is and has been important to a degree which can hardly be overrated. It requires extreme skill, the results of long practice, in detecting the slightest difference in the forms of animals, and it implies some distinct object in view; with these requisites and patience, the breeder has simply to watch for every the smallest approach to the desired end, to select such individuals and pair them with the most suitable forms, and so continue with succeeding generations. In most cases careful selection and the prevention of accidental crosses will be necessary for several generations, for in new breeds there is a strong tendency to vary and especially to revert to ancestral forms: but in every succeeding generation less care will be requisite for the breed will become

did not mean to imply that such varieties are freed from the action of natural selection, but merely that a new form may appear without *summation* of new characters. Professor Henslow is apparently unaware that the above passage is omitted in the second edition of *Var. under Dom.*, II. p. 260.

truer; until ultimately only an occasional individual will require to be separated or destroyed. Horticulturalists in raising seeds regularly practise this, and call it "roguing," or destroying the "rogues" or false varieties. There is another and less efficient means of selection amongst animals: namely repeatedly procuring males with some desirable qualities, and allowing them and their offspring to breed freely together; and this in the course of time will affect the whole lot. These principles of selection have been *methodically* followed for scarcely a century; but their high importance is shown by the practical results, and is admitted in the writings of the most celebrated agriculturalists and horticulturalists;—I need only name Anderson, Marshall, Bakewell, Coke, Western, Sebright and Knight.

Even in well-established breeds the individuals of which to an unpractised eye would appear absolutely similar, which would give, it might have been thought, no scope to selection, the whole appearance of the animal has been changed in a few years (as in the case of Lord Western's sheep), so that practised agriculturalists could scarcely credit that a change had not been effected by a cross with other breeds. Breeders both of plants and animals frequently give their means of selection greater scope, by crossing different breeds and selecting the offspring; but we shall have to recur to this subject again.

The external conditions will doubtless influence and modify the results of the most careful selection; it has been found impossible to prevent certain breeds of cattle from degenerating on mountain pastures; it would probably be impossible to keep the plumage of the wild-duck in the domesticated race; in certain soils, no care has been sufficient to raise cauliflower seed true to its character; and so in many other cases. But with patience it is wonderful what man has effected. He has selected and therefore in one sense made one breed of horses to race and another to pull; he has made sheep with fleeces good for carpets and other sheep good for broadcloth; he has, in the same sense, made one dog to find game and give him notice when found, and another dog to fetch him the game when killed; he has made by selection the fat to lie mixed with the meat in one breed and in another to accumulate in the bowels for the tallow-chandler[1]; he has made the legs of one breed of pigeons long, and the beak of another so short, that it can hardly feed itself; he has previously determined how the feathers on a bird's body shall be coloured, and how the petals of many flowers shall be streaked or fringed, and has given prizes for complete success;—by selection, he has made the leaves of one variety and the flower-buds of another variety of the cabbage good to eat, at different seasons of the year; and thus has he acted on endless varieties. I do not wish to affirm that the long- and short-wooled sheep, or that the pointer and retriever, or that the cabbage and cauliflower have certainly descended from one and the same aboriginal wild stock; if they have not so descended, though it lessens what man has effected, a large result must be left unquestioned.

In saying as I have done that man makes a breed, let it not be confounded with saying that man makes the individuals, which are given by nature with certain desirable qualities; man only adds together and makes a permanent gift of nature's bounties. In several cases, indeed, for instance in the "Ancon" sheep, valuable from not getting over fences, and in the turnspit dog, man has probably only prevented crossing; but in many cases we positively

[1] See the Essay of 1842, p. 3.

know that he has gone on selecting, and taking advantage of successive small variations.

Selection[1] has been *methodically* followed, as I have said, for barely a century; but it cannot be doubted that occasionally it has been practised from the remotest ages, in those animals completely under the dominion of man. In the earliest chapters of the Bible there are rules given for influencing the colours of breeds, and black and white sheep are spoken of as separated. In the time of Pliny the barbarians of Europe and Asia endeavoured by cross-breeding with a wild stock to improve the races of their dogs and horses. The savages of Guyana now do so with their dogs: such care shows at least that the characters of individual animals were attended to. In the rudest times of English history, there were laws to prevent the exportation of fine animals of established breeds, and in the case of horses, in Henry VIII's time, laws for the destruction of all horses under a certain size. In one of the oldest numbers of the *Phil. Transactions*, there are rules for selecting and improving the breeds of sheep. Sir H. Bunbury, in 1660, has given rules for selecting the finest seedling plants, with as much precision as the best recent horticulturalist could. Even in the most savage and rude nations, in the wars and famines which so frequently occur, the most useful of their animals would be preserved: the value set upon animals by savages is shown by the inhabitants of Tierra del Fuego devouring their old women before their dogs, which as they asserted are useful in otter-hunting[2]: who can doubt but that in every case of famine and war, the best otter-hunters would be preserved, and therefore in fact selected for breeding. As the offspring so obviously

[1] See *Origin*, Ed. i. p. 33, vi. p. 38. The evidence is given in the present Essay rather more fully than in the *Origin*.

[2] *Journal of Researches*, Ed. 1860, p. 214. "Doggies catch otters, old women no.''

take after their parents, and as we have seen that savages take pains in crossing their dogs and horses with wild stocks, we may even conclude as probable that they would sometimes pair the most useful of their animals and keep their offspring separate. As different races of men require and admire different qualities in their domesticated animals, each would thus slowly, though unconsciously, be selecting a different breed. As Pallas has remarked, who can doubt but that the ancient Russian would esteem and endeavour to preserve those sheep in his flocks which had the thickest coats. This kind of insensible selection by which new breeds are not selected and kept separate, but a peculiar character is slowly given to the whole mass of the breed, by often saving the life of animals with certain characteristics, we may feel nearly sure, from what we see has been done by the more direct method of separate selection within the last 50 years in England, would in the course of some thousand years produce a marked effect.

Crossing Breeds.

When once two or more races are formed, or if more than one race, or species fertile *inter se*, originally existed in a wild state, their crossing becomes a most copious source of new races[1]. When two well-marked races are crossed the offspring in the first generation take more or less after either parent or are quite intermediate between them, or rarely assume characters in some degree new. In the second and several succeeding generations, the offspring are generally found to

[1] The effects of crossing is much more strongly stated here than in the *Origin*. See Ed. i. p. 20, vi. p. 23, where indeed the opposite point of view is given. His change of opinion may be due to his work on pigeons. The whole of the discussion on crossing corresponds to Chapter VIII of the *Origin*, Ed. i. rather than to anything in the earlier part of the book.

vary exceedingly, one compared with another, and many revert nearly to their ancestral forms. This greater variability in succeeding generations seems analogous to the breaking or variability of organic beings after having been bred for some generations under domestication[1]. So marked is this variability in cross-bred descendants, that Pallas and some other naturalists have supposed that all variation is due to an original cross; but I conceive that the history of the potato, Dahlia, Scotch Rose, the guinea-pig, and of many trees in this country, where only one species of the genus exists, clearly shows that a species may vary where there can have been no crossing. Owing to this variability and tendency to reversion in cross-bred beings, much careful selection is requisite to make intermediate or new permanent races: nevertheless crossing has been a most powerful engine, especially with plants, where means of propagation exist by which the cross-bred varieties can be secured without incurring the risk of fresh variation from seminal propagation: with animals the most skilful agriculturalists now greatly prefer careful selection from a well-established breed, rather than from uncertain cross-bred stocks.

Although intermediate and new races may be formed by the mingling of others, yet if the two races are allowed to mingle quite freely, so that none of either parent race remain pure, then, especially if the parent races are not widely different, they will slowly blend together, and the two races will be destroyed, and one mongrel race left in its place. This will of course happen in a shorter time, if one

[1] The parallelism between the effects of a cross and the effects of conditions is given from a different point of view in the *Origin*, Ed. i. p. 266, vi. p. 391. See the experimental evidence for this important principle in the author's work on *Cross and Self-Fertilisation*. Professor Bateson has suggested that the experiments should be repeated with gametically pure plants.

of the parent races exists in greater number than the other. We see the effect of this mingling, in the manner in which the aboriginal breeds of dogs and pigs in the Oceanic Islands and the many breeds of our domestic animals introduced into S. America, have all been lost and absorbed in a mongrel race. It is probably owing to the freedom of crossing, that, in uncivilised countries, where inclosures do not exist, we seldom meet with more than one race of a species: it is only in enclosed countries, where the inhabitants do not migrate, and have conveniences for separating the several kinds of domestic animals, that we meet with a multitude of races. Even in civilised countries, want of care for a few years has been found to destroy the good results of far longer periods of selection and separation.

This power of crossing will affect the races of all *terrestrial* animals; for all terrestrial animals require for their reproduction the union of two individuals. Amongst plants, races will not cross and blend together with so much freedom as in terrestrial animals; but this crossing takes place through various curious contrivances to a surprising extent. In fact such contrivances exist in so very many hermaphrodite flowers by which an occasional cross may take place, that I cannot avoid suspecting (with Mr Knight) that the reproductive action requires, at *intervals*, the concurrence of distinct individuals[1]. Most breeders of plants and animals are firmly convinced that benefit is derived from an occasional cross, not with another race, but with another family of the same race; and that, on the other hand, injurious consequences follow from long-continued close interbreeding in the same

[1] The so-called Knight-Darwin Law is often misunderstood. See Goebel in *Darwin and Modern Science*, 1909, p. 419; also F. Darwin, *Nature*, Oct. 27, 1898.

family. Of marine animals, many more, than was till lately believed, have their sexes on separate individuals; and where they are hermaphrodite, there seems very generally to be means through the water of one individual occasionally impregnating another: if individual animals can singly propagate themselves for perpetuity, it is unaccountable that no terrestrial animal, where the means of observation are more obvious, should be in this predicament of singly perpetuating its kind. I conclude, then, that races of most animals and plants, when unconfined in the same country, would tend to blend together.

Whether our domestic races have descended from one or more wild stocks.

Several naturalists, of whom Pallas[1] regarding animals, and Humboldt regarding certain plants, were the first, believe that the breeds of many of our domestic animals such as of the horse, pig, dog, sheep, pigeon, and poultry, and of our plants have descended from more than one aboriginal form. They leave it doubtful, whether such forms are to be considered wild races, or true species, whose offspring are fertile when crossed *inter se.* The main arguments for this view consist, firstly, of the great difference between such breeds, as the Race- and Cart-Horse, or the Greyhound and Bull-dog, and of our ignorance of the steps or stages through which these could have passed from a common parent; and secondly that in the most ancient historical periods, breeds resembling some of those at present most different, existed in different countries. The wolves of N. America and of Siberia are thought to be different species; and

[1] Pallas' theory is discussed in the *Origin*, Ed. i. pp. 253, 254, vi. p. 374.

it has been remarked that the dogs belonging to the savages in these two countries resemble the wolves of the same country; and therefore that they have probably descended from two different wild stocks. In the same manner, these naturalists believe that the horse of Arabia and of Europe have probably descended from two wild stocks both apparently now extinct. I do not think the assumed fertility of these wild stocks any very great difficulty on this view; for although in animals the offspring of most cross-bred species are infertile, it is not always remembered that the experiment is very seldom fairly tried, except when two near species *both* breed freely (which does not readily happen, as we shall hereafter see) when under the dominion of man. Moreover in the case of the China[1] and common goose, the canary and siskin, the hybrids breed freely; in other cases the offspring from hybrids crossed with either pure parent are fertile, as is practically taken advantage of with the yak and cow; as far as the analogy of plants serves, it is impossible to deny that some species are quite fertile *inter se*; but to this subject we shall recur.

On the other hand, the upholders of the view that the several breeds of dogs, horses, &c., &c., have descended each from one stock, may aver that their view removes all *difficulty about fertility*, and that the main argument from the high antiquity of different breeds, somewhat similar to the present breeds, is worth little without knowing the date of the domestication of such animals, which is far from being the case. They may also with more weight aver that, knowing that organic beings under domestication do vary in some degree, the argument from the great difference between certain breeds is

[1] See Darwin's paper on the fertility of hybrids from the common and Chinese goose in *Nature*, Jan. 1, 1880.

worth nothing, without we know the limits of variation during a long course of time, which is far from the case. They may argue that almost every county in England, and in many districts of other countries, for instance in India, there are slightly different breeds of the domestic animals; and that it is opposed to all that we know of the distribution of wild animals to suppose that these have descended from so many different wild races or species: if so, they may argue, is it not probable that countries quite separate and exposed to different climates would have breeds not slightly, but considerably, different? Taking the most favourable case, on both sides, namely that of the dog; they might urge that such breeds as the bull-dog and turnspit have been reared by man, from the ascertained fact that strictly analogous breeds (namely the Niata ox and Ancon sheep) in other quadrupeds have thus originated. Again they may say, seeing what training and careful selection has effected for the greyhound, and seeing how absolutely unfit the Italian greyhound is to maintain itself in a state of nature, is it not probable that at least all greyhounds,—from the rough deerhound, the smooth Persian, the common English, to the Italian,—have descended from one stock[1]? If so, is it so improbable that the deerhound and long-legged shepherd dog have so descended? If we admit this, and give up the bull-dog, we can hardly dispute the probable common descent of the other breeds.

The evidence is so conjectural and balanced on both sides that at present I conceive that no one can decide: for my own part, I lean to the probability of most of our domestic animals having descended from more than one wild stock; though from the arguments last advanced and from reflecting on the slow though inevitable effect of

[1] *Origin*, Ed. i. p. 19, vi. p. 22.

different races of mankind, under different circumstances, saving the lives of and therefore selecting the individuals most useful to them, I cannot doubt but that one class of naturalists have much overrated the probable number of the aboriginal wild stocks. As far as we admit the difference of our races ⟨to be⟩ due to the differences of their original stocks, so much must we give up of the amount of variation produced under domestication. But this appears to me unimportant, for we certainly know in some few cases, for instance in the Dahlia, and potato, and rabbit, that a great number of varieties have proceeded from one stock; and, in many of our domestic races, we know that man, by slowly selecting and by taking advantage of sudden sports, has considerably modified old races and produced new ones. Whether we consider our races as the descendants of one or several wild stocks, we are in far the greater number of cases equally ignorant what these stocks were.

Limits to Variation in degree and kind.

Man's power in making races depends, in the first instance, on the stock on which he works being variable; but his labours are modified and limited, as we have seen, by the direct effects of the external conditions,—by the deficient or imperfect hereditariness of new peculiarities,—and by the tendency to continual variation and especially to reversion to ancestral forms. If the stock is not variable under domestication, of course he can do nothing; and it appears that species differ considerably in this tendency to variation, in the same way as even sub-varieties from the same variety differ greatly in this respect, and transmit to their offspring this difference in tendency. Whether the absence of a tendency to vary is an unalterable quality in certain species, or depends on some deficient condition of the particular state of domestication to which they are exposed, there is no evidence. When the organization is rendered variable, or plastic, as I have expressed it, under domestication, different parts of the frame vary more or less in different species: thus in the breeds of cattle it has been remarked that the horns are the most constant or least variable character, for these often remain constant, whilst the colour, size, proportions of the body, tendency to fatten &c., vary; in sheep, I believe, the horns are much more variable. As a general rule the less important parts of the organization seem to vary most, but I think there is sufficient evidence that every part occasionally varies in a slight degree. Even when man has the primary requisite variability he is necessarily checked by the health and life of the stock he is working on: thus he has already made pigeons with such small beaks that they can hardly eat and will not rear their own young; he has made families of sheep with so strong a tendency to early maturity and to fatten, that in certain pastures they cannot live from their extreme liability to inflammation; he has made (*i.e.* selected) sub-varieties of plants with a tendency to such early growth that they are frequently killed by the spring frosts; he has made a breed of cows having calves with such large hinder quarters that they are born with great difficulty, often to the death of their mothers[1]; the breeders were compelled to remedy this by the selection of a breeding stock with smaller hinder quarters; in such a case, however, it is possible by long patience and great loss, a remedy might have been found in selecting cows capable of giving birth to calves with large hinder quarters, for in human kind there ⟨are⟩ no doubt hereditary bad and

[1] *Var. under Dom.*, Ed. ii. vol. II. p. 211.

good confinements. Besides the limits already specified, there can be little doubt that the variation of different parts of the frame are connected together by many laws[1]: thus the two sides of the body, in health and disease, seem almost always to vary together: it has been asserted by breeders that if the head is much elongated, the bones of the extremities will likewise be so; in seedling-apples large leaves and fruit generally go together, and serve the horticulturalist as some guide in his selection; we can here see the reason, as the fruit is only a metamorphosed leaf. In animals the teeth and hair seem connected, for the hairless Chinese dog is almost toothless. Breeders believe that one part of the frame or function being increased causes other parts to decrease: they dislike great horns and great bones as so much flesh lost; in hornless breeds of cattle certain bones of the head become more developed: it is said that fat accumulating in one part checks its accumulation in another, and likewise checks the action of the udder. The whole organization is so connected that it is probable there are many conditions determining the variation of each part, and causing other parts to vary with it; and man in making new races must be limited and ruled by all such laws.

In what consists Domestication.

In this chapter we have treated of variation under domestication, and it now remains to consider in what does this power of domestication consist[2], a subject of considerable difficulty. Observing that organic beings of almost every class, in all climates, countries, and times, have varied when long bred

[1] This discussion corresponds to the *Origin*, Ed. i. pp. 11 and 143, vi. pp. 13 and 177.

[2] See *Origin*, Ed. i. p. 7, vi. p. 7.

under domestication, we must conclude that the influence is of some very general nature[1]. Mr Knight alone, as far as I know, has tried to define it; he believes it consists of an excess of food, together with transport to a more genial climate, or protection from its severities. I think we cannot admit this latter proposition, for we know how many vegetable products, aborigines of this country, here vary, when cultivated without any protection from the weather; and some of our variable trees, as apricots, peaches, have undoubtedly been derived from a more genial climate. There appears to be much more truth in the doctrine of excess of food being the cause, though I much doubt whether this is the sole cause, although it may well be requisite for the kind of variation desired by man, namely increase of size and vigour. No doubt horticulturalists, when they wish to raise new seedlings, often pluck off all the flower-buds, except a few, or remove the whole during one season, so that a great stock of nutriment may be thrown into the flowers which are to seed. When plants are transported from high-lands, forests, marshes, heaths, into our gardens and greenhouses, there must be a considerable change of food, but it would be hard to prove that there was in every case an excess of the kind proper to the plant. If it be an excess of food, compared with that which the being obtained in its natural state[2], the effects continue for an improbably long time; during how many ages has

[1] ⟨Note in the original.⟩ "Isidore G. St Hilaire insists that breeding in captivity essential element. Schleiden on alkalies. ⟨See *Var. under Dom.*, Ed. ii. vol. II. p. 244, note 10.⟩ What is it in domestication which causes variation?"

[2] ⟨Note in the original.⟩ "It appears that slight changes of condition ⟨are⟩ good for health; that more change affects the generative system, so that variation results in the offspring; that still more change checks or destroys fertility not of the offspring." Compare the *Origin*, Ed. i. p. 9, vi. p. 11. What the meaning of "not of the offspring" may be is not clear.

wheat been cultivated, and cattle and sheep reclaimed, and we cannot suppose their *amount* of food has gone on increasing, nevertheless these are amongst the most variable of our domestic productions. It has been remarked (Marshall) that some of the most highly kept breeds of sheep and cattle are truer or less variable than the straggling animals of the poor, which subsist on commons, and pick up a bare subsistence[1]. In the case of forest-trees raised in nurseries, which vary more than the same trees do in their aboriginal forests, the cause would seem simply to lie in their not having to struggle against other trees and weeds, which in their natural state doubtless would limit the conditions of their existence. It appears to me that the power of domestication resolves itself into the accumulated effects of a change of all or some of the natural conditions of the life of the species, often associated with excess of food. These conditions moreover, I may add, can seldom remain, owing to the mutability of the affairs, habits, migrations, and knowledge of man, for very long periods the same. I am the more inclined to come to this conclusion from finding, as we shall hereafter show, that changes of the natural conditions of existence seem peculiarly to affect the action of the reproductive system[2]. As we see that hybrids and mongrels, after the first generation, are apt to vary much, we may at least conclude that variability does not altogether depend on excess of food.

After these views, it may be asked how it comes

[1] In the *Origin*, Ed. i. p. 41, vi. p. 46 the question is differently treated; it is pointed out that a large stock of individuals gives a better chance of available variations occurring. Darwin quotes from Marshall that sheep in small lots can never be improved. This comes from Marshall's *Review of the Reports to the Board of Agriculture*, 1808, p. 406. In this Essay the name Marshall occurs in the margin. Probably this refers to *loc. cit.* p. 200, where unshepherded sheep in many parts of England are said to be similar owing to mixed breeding not being avoided.

[2] See *Origin*, Ed. i. p. 8, vi. p. 8.

that certain animals and plants, which have been domesticated for a considerable length of time, and transported from very different conditions of existence, have not varied much, or scarcely at all; for instance, the ass, peacock, guinea-fowl, asparagus, Jerusalem artichoke[1]. I have already said that probably different species, like different sub-varieties, possess different degrees of tendency to vary; but I am inclined to attribute in these cases the want of numerous races less to want of variability than to selection not having been practised on them. No one will take the pains to select without some corresponding object, either of use or amusement; the individuals raised must be tolerably numerous, and not so precious, but that he may freely destroy those not answering to his wishes. If guinea-fowls or peacocks[2] became "fancy" birds, I cannot doubt that after some generations several breeds would be raised. Asses have not been worked on from mere neglect; but they differ in *some* degree in different countries. The insensible selection, due to different races of mankind preserving those individuals most useful to them in their different circumstances, will apply only to the oldest and most widely domesticated animals. In the case of plants, we must put entirely out of the case those exclusively (or almost so) propagated by cuttings, layers or tubers, such as the Jerusalem artichoke and laurel; and if we put on one side plants of little ornament or use, and those which are used at so early a period of their growth that no especial characters signify, as asparagus[3] and seakale, I can think of none long cultivated which have not varied. In no case ought we to expect to find as much variation in a race when it alone has been formed, as when several have been formed,

[1] See *Origin*, Ed. i. p. 42, vi. p. 48.
[2] ⟨Note in the original.⟩ There are white peacocks.
[3] ⟨Note in the original.⟩ There are varieties of asparagus.

for their crossing and recrossing will greatly increase their variability.

Summary of first Chapter.

To sum up this chapter. Races are made under domestication: 1st, by the direct effects of the external conditions to which the species is exposed: 2nd, by the indirect effects of the exposure to new conditions, often aided by excess of food, rendering the organization plastic, and by man's selecting and separately breeding certain individuals, or introducing to his stock selected males, or often preserving with care the life of the individuals best adapted to his purposes: 3rd, by crossing and recrossing races already made, and selecting their offspring. After some generations man may relax his care in selection: for the tendency to vary and to revert to ancestral forms will decrease, so that he will have only occasionally to remove or destroy one of the yearly offspring which departs from its type. Ultimately, with a large stock, the effects of free crossing would keep, even without this care, his breed true. By these means man can produce infinitely numerous races, curiously adapted to ends, both most important and most frivolous; at the same time that the effects of the surrounding conditions, the laws of inheritance, of growth, and of variation, will modify and limit his labours.

CHAPTER II

ON THE VARIATION OF ORGANIC BEINGS IN A WILD STATE; ON THE NATURAL MEANS OF SELECTION; AND ON THE COMPARISON OF DOMESTIC RACES AND TRUE SPECIES

HAVING treated of variation under domestication, we now come to it in a *state of nature.*

Most organic beings in a state of nature vary exceedingly little[1]: I put out of the case variations (as stunted plants &c., and sea-shells in brackish water[2]) which are directly the effect of external agencies and which we do not *know are in the breed*[3], or are *hereditary.* The amount of hereditary variation is very difficult to ascertain, because naturalists (partly from the want of knowledge, and partly from the inherent difficulty of the subject) do not all agree whether certain forms are species or races[4]. Some strongly marked races of plants, comparable with the decided sports of horti-

[1] In Chapter II of the first edition of the *Origin* Darwin insists rather on the presence of variability in a state of nature; see, for instance, p. 45, Ed. vi. p. 53, "I am convinced that the most experienced naturalist would be surprised at the number of the cases of variability...which he could collect on good authority, as I have collected, during a course of years."

[2] See *Origin*, Ed. i. p. 44, vi. p. 52.

[3] ⟨Note in the original.⟩ Here discuss *what is a species*, sterility can most rarely be told when crossed.—Descent from common stock.

[4] ⟨Note in the original.⟩ Give only rule: chain of intermediate forms, and *analogy*; this important. Every Naturalist at first when he gets hold of new variable type is *quite puzzled* to know what to think species and what variations.

culturalists, undoubtedly exist in a state of nature, as is actually known by experiment, for instance in the primrose and cowslip[1], in two so-called species of dandelion, in two of foxglove[2], and I believe in some pines. Lamarck has observed that, as long as we confine our attention to one limited country, there is seldom much difficulty in deciding what forms to call species and what varieties; and that it is when collections flow in from all parts of the world that naturalists often feel at a loss to decide the limit of variation. Undoubtedly so it is, yet amongst British plants (and I may add land shells), which are probably better known than any in the world, the best naturalists differ very greatly in the relative proportions of what they call species and what varieties. In many genera of insects, and shells, and plants, it seems almost hopeless to establish which are which. In the higher classes there are less doubts; though we find considerable difficulty in ascertaining what deserve to be called species amongst foxes and wolves, and in some birds, for instance in the case of the white barn-owl. When specimens are brought from different parts of the world, how often do naturalists dispute this same question, as I found with respect to the birds brought from the Galapagos islands. Yarrell has remarked that the individuals of the same undoubted species of birds, from Europe and N. America, usually present slight, indefinable though perceptible differences. The recognition indeed of

[1] The author had not at this time the knowledge of the meaning of dimorphism.

[2] ⟨Note in original.⟩ Compare feathered heads in very different birds with spines in Echidna and Hedgehog. ⟨In *Variation under Domestication*, Ed. ii. vol. II. p. 317, Darwin calls attention to laced and frizzled breeds occurring in both fowls and pigeons. In the same way a peculiar form of covering occurs in Echidna and the hedgehog.⟩

Plants under very different climate not varying. Digitalis shows jumps ⟨?⟩ in variation, like Laburnum and Orchis case—in fact hostile cases. Variability of sexual characters alike in domestic and wild.

one animal by another of its kind seems to imply some difference. The disposition of wild animals undoubtedly differs. The variation, such as it is, chiefly affects the same parts in wild organisms as in domestic breeds; for instance, the size, colour, and the external and less important parts. In many species the variability of certain organs or qualities is even stated as one of the specific characters: thus, in plants, colour, size, hairiness, the number of the stamens and pistils, and even their presence, the form of the leaves; the size and form of the mandibles of the males of some insects; the length and curvature of the beak in some birds (as in Opetiorynchus) are variable characters in some species and quite fixed in others. I do not perceive that any just distinction can be drawn between this recognised variability of certain parts in many species and the more general variability of the whole frame in domestic races.

Although the amount of variation be exceedingly small in most organic beings in a state of nature, and probably quite wanting (as far as our senses serve) in the majority of cases; yet considering how many animals and plants, taken by mankind from different quarters of the world for the most diverse purposes, have varied under domestication in every country and in every age, I think we may safely conclude that all organic beings with few exceptions, if capable of being domesticated and bred for long periods, would vary. Domestication seems to resolve itself into a change from the natural conditions of the species [generally perhaps including an increase of food]; if this be so, organisms in a state of nature must *occasionally*, in the course of ages, be exposed to analogous influences; for geology clearly shows that many places must, in the course of time, become exposed to the widest range of climatic and other influences; and if such places be isolated, so that new and better adapted organic beings cannot freely emigrate, the old inhabitants will be exposed to new influences, probably far more varied, than man applies under the form of domestication. Although every species no doubt will soon breed up to the full number which the country will support, yet it is easy to conceive that, on an average, some species may receive an increase of food; for the times of dearth may be short, yet enough to kill, and recurrent only at long intervals. All such changes of conditions from geological causes would be exceedingly slow; what effect the slowness might have we are ignorant; under domestication it appears that the effects of change of conditions accumulate, and then break out. Whatever might be the result of these slow geological changes, we may feel sure, from the means of dissemination common in a lesser or greater degree to every organism taken conjointly with the changes of geology, which are steadily (and sometimes suddenly, as when an isthmus at last separates) in progress, that occasionally organisms must suddenly be introduced into new regions, where, if the conditions of existence are not so foreign as to cause its extermination, it will often be propagated under circumstances still more closely analogous to those of domestication; and therefore we expect will evince a tendency to vary. It appears to me quite *inexplicable* if this has never happened; but it can happen very rarely. Let us then suppose that an organism by some chance (which might be hardly repeated in 1000 years) arrives at a modern volcanic island in process of formation and not fully stocked with the most appropriate organisms; the new organism might readily gain a footing, although the external conditions were considerably different from its native ones. The effect of this we might expect would influence in some small degree the size, colour, nature of covering &c., and from inexplicable influences even special parts and organs of the body. But we might further (and this is far more important) expect that the reproductive system would be affected, as under domesticity, and the structure of the offspring rendered in some degree plastic. Hence almost every part of the body would tend to vary from the typical form in slight degrees, and in no determinate way, and therefore *without selection* the free crossing of these small variations (together with the tendency to reversion to the original form) would constantly be counteracting this unsettling effect of the extraneous conditions on the reproductive system. Such, I conceive, would be the unimportant result without selection. And here I must observe that the foregoing remarks are equally applicable to that small and admitted amount of variation which has been observed in some organisms in a state of nature; as well as to the above hypothetical variation consequent on changes of condition.

Let us now suppose a Being[1] with penetration sufficient to perceive differences in the outer and innermost organization quite imperceptible to man, and with forethought extending over future centuries to watch with unerring care and select for any object the offspring of an organism produced under the foregoing circumstances; I can see no conceivable reason why he could not form a new race (or several were he to separate the stock of the original organism and work on several islands) adapted to new ends. As we assume his discrimination, and his forethought, and his steadiness of object, to be incomparably greater that those qualities in man, so we may suppose the beauty and complications of the adaptations of the new races and their differences from the original stock to be greater than in the domestic races produced by man's agency: the

[1] A corresponding passage occurs in *Origin*, Ed. i. p. 83, vi. p. 101, where however Nature takes the place of the selecting Being.

ground-work of his labours we may aid by supposing that the external conditions of the volcanic island, from its continued emergence and the occasional introduction of new immigrants, vary; and thus to act on the reproductive system of the organism, on which he is at work, and so keep its organization somewhat plastic. With time enough, such a Being might rationally (without some unknown law opposed him) aim at almost any result.

For instance, let this imaginary Being wish, from seeing a plant growing on the decaying matter in a forest and choked by other plants, to give it power of growing on the rotten stems of trees, he would commence selecting every seedling whose berries were in the smallest degree more attractive to tree-frequenting birds, so as to cause a proper dissemination of the seeds, and at the same time he would select those plants which had in the slightest degree more and more power of drawing nutriment from rotten wood; and he would destroy all other seedlings with less of this power. He might thus, in the course of century after century, hope to make the plant by degrees grow on rotten wood, even high up on trees, wherever birds dropped the non-digested seeds. He might then, if the organization of the plant was plastic, attempt by continued selection of chance seedlings to make it grow on less and less rotten wood, till it would grow on sound wood[1]. Supposing again, during these changes the plant failed to seed quite freely from non-impregnation, he might begin selecting seedlings with a little sweeter ⟨or⟩ differently tasted honey or pollen, to tempt insects to visit the flowers regularly: having effected this, he might wish, if it profited the plant, to render abortive the stamens and pistils in different flowers, which he could do by continued selection. By such

[1] The mistletoe is used as an illustration in *Origin*, Ed. i. p. 3, vi. p. 3, but with less detail.

steps he might aim at making a plant as wonderfully related to other organic beings as is the mistletoe, whose existence absolutely depends on certain insects for impregnation, certain birds for transportal, and certain trees for growth. Furthermore, if the insect which had been induced regularly to visit this hypothetical plant profited much by it, our same Being might wish by selection to modify by gradual selection the insect's structure, so as to facilitate its obtaining the honey or pollen: in this manner he might adapt the insect (always presupposing its organization to be in some degree plastic) to the flower, and the impregnation of the flower to the insect; as is the case with many bees and many plants.

Seeing what blind capricious man has actually effected by selection during the few last years, and what in a ruder state he has probably effected without any systematic plan during the last few thousand years, he will be a bold person who will positively put limits to what the supposed Being could effect during whole geological periods. In accordance with the plan by which this universe seems governed by the Creator, let us consider whether there exists any *secondary* means in the economy of nature by which the process of selection could go on adapting, nicely and wonderfully, organisms, if in ever so small a degree plastic, to diverse ends. I believe such secondary means do exist[1].

Natural means of Selection[2].

De Candolle, in an eloquent passage, has declared that all nature is at war, one organism with another,

[1] ⟨Note in original.⟩ The selection, in cases where adult lives only few hours as Ephemera, must fall on larva—curious speculation of the effect ⟨which⟩ changes in it would bring in parent.

[2] This section forms part of the joint paper by Darwin and Wallace read before the Linnean Society on July 1, 1858.

or with external nature. Seeing the contented face of nature, this may at first be well doubted; but reflection will inevitably prove it is too true. The war, however, is not constant, but only recurrent in a slight degree at short periods and more severely at occasional more distant periods; and hence its effects are easily overlooked. It is the doctrine of Malthus applied in most cases with ten-fold force. As in every climate there are seasons for each of its inhabitants of greater and less abundance, so all annually breed; and the moral restraint, which in some small degree checks the increase of mankind, is entirely lost. Even slow-breeding mankind has doubled in 25 years[1], and if he could increase his food with greater ease, he would double in less time. But for animals, without artificial means, *on an average* the amount of food for each species must be constant; whereas the increase of all organisms tends to be geometrical, and in a vast majority of cases at an enormous ratio. Suppose in a certain spot there are eight pairs of [robins] birds, and that *only* four pairs of them annually (including double hatches) rear only four young; and that these go on rearing their young at the same rate: then at the end of seven years (a short life, excluding violent deaths, for any birds) there will be 2048 robins, instead of the original sixteen; as this increase is quite impossible, so we must conclude either that robins do not rear nearly half their young or that the average life of a robin when reared is from accident not nearly seven years. Both checks probably concur. The same kind of calculation applied to all vegetables and animals produces results either more or less striking, but in scarcely a single instance less striking than in man[2].

Many practical illustrations of this rapid tendency

[1] Occurs in *Origin*, Ed. i. p. 64, vi. p. 79.
[2] Corresponds approximately with *Origin*, Ed. i. pp. 64—65, vi. p. 80.

to increase are on record, namely during peculiar seasons, in the extraordinary increase of certain animals, for instance during the years 1826 to 1828, in La Plata, when from drought, some millions of cattle perished, the whole country *swarmed* with innumerable mice: now I think it cannot be doubted that during the breeding season all the mice (with the exception of a few males or females in excess) ordinarily pair; and therefore that this astounding increase during three years must be attributed to a greater than usual number surviving the first year, and then breeding, and so on, till the third year, when their numbers were brought down to their usual limits on the return of wet weather. Where man has introduced plants and animals into a new country favourable to them, there are many accounts in how surprisingly few years the whole country has become stocked with them. This increase would necessarily stop as soon as the country was fully stocked; and yet we have every reason to believe from what is known of wild animals that *all* would pair in the spring. In the majority of cases it is most difficult to imagine where the check falls, generally no doubt on the seeds, eggs, and young; but when we remember how impossible even in mankind (so much better known than any other animal) it is to infer from repeated casual observations what the average of life is, or to discover how different the percentage of deaths to the births in different countries, we ought to feel no legitimate surprise at not seeing where the check falls in animals and plants. It should always be remembered that in most cases the checks are yearly recurrent in a small regular degree, and in an extreme degree during occasionally unusually cold, hot, dry, or wet years, according to the constitution of the being in question. Lighten any check in the smallest degree, and the geometrical power of increase in every organism will instantly increase the average numbers of the favoured species. Nature may be compared to a surface, on which rest ten thousand sharp wedges touching each other and driven inwards by incessant blows[1]. Fully to realise these views much reflection is requisite; Malthus on man should be studied; and all such cases as those of the mice in La Plata, of the cattle and horses when first turned out in S. America, of the robins by our calculation, &c., should be well considered: reflect on the enormous multiplying power *inherent and annually in action* in all animals; reflect on the countless seeds scattered by a hundred ingenious contrivances, year after year, over the whole face of the land; and yet we have every reason to suppose that the average percentage of every one of the inhabitants of a country will *ordinarily* remain constant. Finally, let it be borne in mind that this average number of individuals (the external conditions remaining the same) in each country is kept up by recurrent struggles against other species or against external nature (as on the borders of the arctic regions[2], where the cold checks life); and that ordinarily each individual of each species holds its place, either by its own struggle and capacity of acquiring nourishment in some period (from the egg upwards) of its life, or by the struggle of its parents (in short lived organisms, when the main check occurs at long intervals) against and compared with other individuals of the *same* or *different* species.

But let the external conditions of a country change; if in a small degree, the relative proportions of the inhabitants will in most cases simply be

[1] This simile occurs in *Origin*, Ed. i. p. 67, not in the later editions.

[2] ⟨Note in the original.⟩ In case like mistletoe, it may be asked why not more species, no other species interferes; answer almost sufficient, same causes which check the multiplication of individuals.

slightly changed; but let the number of inhabitants be small, as in an island[1], and free access to it from other countries be circumscribed; and let the change of condition continue progressing (forming new stations); in such case the original inhabitants must cease to be so perfectly adapted to the changed conditions as they originally were. It has been shown that probably such changes of external conditions would, from acting on the reproductive system, cause the organization of the beings most affected to become, as under domestication, plastic. Now can it be doubted from the struggle each individual (or its parents) has to obtain subsistence that any minute variation in structure, habits, or instincts, adapting that individual better to the new conditions, would tell upon its vigour and health? In the struggle it would have a better *chance* of surviving, and those of its offspring which inherited the variation, let it be ever so slight, would have a better *chance* to survive. Yearly more are bred than can survive; the smallest grain in the balance, in the long run, must tell on which death shall fall, and which shall survive[2]. Let this work of selection, on the one hand, and death on the other, go on for a thousand generations; who would pretend to affirm that it would produce no effect, when we remember what in a few years Bakewell effected in cattle and Western in sheep, by this identical principle of selection.

To give an imaginary example, from changes in progress on an island, let the organization[3] of a canine animal become slightly plastic, which animal preyed chiefly on rabbits, but sometimes on hares; let these same changes cause the number of rabbits

[1] See *Origin*, Ed. i. pp. 104, 292, vi. pp. 127, 429.

[2] Recognition of the importance of minute differences in the struggle occurs in the Essay of 1842, p. 8 note 3.

[3] See *Origin*, Ed. i. p. 90, vi. p. 110.

very slowly to decrease and the number of hares to increase; the effect of this would be that the fox or dog would be driven to try to catch more hares, and his numbers would tend to decrease; his organization, however, being slightly plastic, those individuals with the lightest forms, longest limbs, and best eyesight (though perhaps with less cunning or scent) would be slightly favoured, let the difference be ever so small, and would tend to live longer and to survive during that time of the year when food was shortest; they would also rear more young, which young would tend to inherit these slight peculiarities. The less fleet ones would be rigidly destroyed. I can see no more reason to doubt but that these causes in a thousand generations would produce a marked effect, and adapt the form of the fox to catching hares instead of rabbits, than that greyhounds can be improved by selection and careful breeding. So would it be with plants under similar circumstances; if the number of individuals of a species with plumed seeds could be increased by greater powers of dissemination within its own area (that is if the check to increase fell chiefly on the seeds), those seeds which were provided with ever so little more down, or with a plume placed so as to be slightly more acted on by the winds, would in the long run tend to be most disseminated; and hence a greater number of seeds thus formed would germinate, and would tend to produce plants inheriting this slightly better adapted down.

Besides this natural means of selection, by which those individuals are preserved, whether in their egg or seed or in their mature state, which are best adapted to the place they fill in nature, there is a second agency at work in most bisexual animals tending to produce the same effect, namely the struggle of the males for the females. These struggles are generally decided by the law of battle;

but in the case of birds, apparently, by the charms of their song[1], by their beauty or their power of courtship, as in the dancing rock-thrush of Guiana. Even in the animals which pair there seems to be an excess of males which would aid in causing a struggle: in the polygamous animals[2], however, as in deer, oxen, poultry, we might expect there would be severest struggle: is it not in the polygamous animals that the males are best formed for mutual war? The most vigorous males, implying perfect adaptation, must generally gain the victory in their several contests. This kind of selection, however, is less rigorous than the other; it does not require the death of the less successful, but gives to them fewer descendants. This struggle falls, moreover, at a time of year when food is generally abundant, and perhaps the effect chiefly produced would be the alteration of sexual characters, and the selection of individual forms, no way related to their power of obtaining food, or of defending themselves from their natural enemies, but of fighting one with another. This natural struggle amongst the males may be compared in effect, but in a less degree, to that produced by those agriculturalists who pay less attention to the careful selection of all the young animals which they breed and more to the occasional use of a choice male[3].

[1] These two forms of sexual selection are given in *Origin*, Ed. i. p. 87, vi. p. 107. The Guiana rock-thrush is given as an example of bloodless competition.

[2] ⟨Note in original.⟩ Seals? Pennant about battles of seals.

[3] In the Linnean paper of July 1, 1858 the final word is *mate*: but the context shows that it should be *male*; it is moreover clearly so written in the MS.

Differences between "Races" and "Species":—first, in their trueness or variability.

Races[1] produced by these natural means of selection[2] we may expect would differ in some respects from those produced by man. Man selects chiefly by the eye, and is not able to perceive the course of every vessel and nerve, or the form of the bones, or whether the internal structure corresponds to the outside shape. He[3] is unable to select shades of constitutional differences, and by the protection he affords and his endeavours to keep his property alive, in whatever country he lives, he checks, as much as lies in his power, the selecting action of nature, which will, however, go on to a lesser degree with all living things, even if their length of life is not determined by their own powers of endurance. He has bad judgment, is capricious, he does not, or his successors do not, wish to select for the same exact end for hundreds of generations. He cannot always suit the selected form to the properest conditions; nor does he keep those conditions uniform: he selects that which is useful to him, not that best adapted to those conditions in which each variety is placed by him: he selects a small dog, but feeds it highly; he selects a long-backed dog, but does not exercise it in any peculiar manner, at least not during every generation. He seldom allows the most vigorous males to struggle for themselves and propagate, but picks out such as he possesses, or such as he prefers, and not necessarily those best adapted to the existing conditions. Every agriculturalist and breeder knows how difficult it is to prevent an occasional cross with another breed.

[1] In the *Origin* the author would here have used the word *variety*.

[2] The whole of p. 94 and 15 lines of p. 95 are, in the MS., marked through in pencil with vertical lines, beginning at "Races produced, &c." and ending with "to these conditions."

[3] See *Origin*, Ed. i. p. 83, vi. p. 102.

He often grudges to destroy an individual which departs considerably from the required type. He often begins his selection by a form or sport considerably departing from the parent form. Very differently does the natural law of selection act; the varieties selected differ only slightly from the parent forms[1]; the conditions are constant for long periods and change slowly; rarely can there be a cross; the selection is rigid and unfailing, and continued through many generations; a selection can *never be made* without the form be *better* adapted to the conditions than the parent form; the selecting power goes on without caprice, and steadily for thousands of years adapting the form to these conditions. The selecting power is not deceived by external appearances, it tries the being during its whole life; and if less well ⟨?⟩ adapted than its *congeners*, without fail it is destroyed; every part of its structure is thus scrutinised and proved good towards the place in nature which it occupies.

We have every reason to believe that in proportion to the number of generations that a domestic race is kept free from crosses, and to the care employed in continued steady selection with one end in view, and to the care in not placing the variety in conditions unsuited to it; in such proportion does the new race become "true" or subject to little variation[2]. How incomparably "truer" then would a race produced by the above rigid, steady, natural means of selection, excellently trained and perfectly adapted to its conditions, free from stains of blood or crosses, and continued during thousands of years, be compared with one produced by the feeble, capri-

[1] In the present Essay there is some evidence that the author attributed more to *sports* than was afterwards the case: but the above passage points the other way. It must always be remembered that many of the minute differences, now considered small mutations, are the small variations on which Darwin conceived selection to act.

[2] See *Var. under Dom.*, Ed. ii. vol. II. p. 230.

cious, misdirected and ill-adapted selection of man. Those races of domestic animals produced by savages, partly by the inevitable conditions of their life, and partly unintentionally by their greater care of the individuals most valuable to them, would probably approach closest to the character of a species; and I believe this is the case. Now the characteristic mark of a species, next, if not equal in importance to its sterility when crossed with another species, and indeed almost the only other character (without we beg the question and affirm the essence of a species, is its not having descended from a parent common to any other form), is the similarity of the individuals composing the species, or in the language of agriculturalists their "trueness."

Difference between "Races" and "Species" in fertility when crossed.

The sterility of species, or of their offspring, when crossed has, however, received more attention than the uniformity in character of the individuals composing the species. It is exceedingly natural that such sterility[1] should have been long thought the certain characteristic of species. For it is obvious that if the allied different forms which we meet with in the same country could cross together, instead of finding a number of distinct species, we should have a confused and blending series. The fact however of a perfect gradation in the degree of sterility between species, and the circumstance of some species most closely allied (for instance many species of crocus and European heaths) refusing

[1] ⟨Note in the original.⟩ If domestic animals are descended from several species and *become* fertile *inter se*, then one can see they gain fertility by becoming adapted to new conditions and certainly domestic animals can withstand changes of climate without loss of fertility in an astonishing manner.

to breed together, whereas other species, widely different, and even belonging to distinct genera, as the fowl and the peacock, pheasant and grouse[1], Azalea and Rhododendron, Thuja and Juniperus, breeding together ought to have caused a doubt whether the sterility did not depend on other causes, distinct from a law, coincident with their creation. I may here remark that the fact whether one species will or will not breed with another is far less important than the sterility of the offspring when produced; for even some domestic races differ so greatly in size (as the great stag-greyhound and lap-dog, or cart-horse and Burmese ponies) that union is nearly impossible; and what is less generally known is, that in plants Kölreuter has shown by hundreds of experiments that the pollen of one species will fecundate the germen of another species, whereas the pollen of this latter will never act on the germen of the former; so that the simple fact of mutual impregnation certainly has no relation whatever to the distinctness in creation of the two forms. When two species are attempted to be crossed which are so distantly allied that offspring are never produced, it has been observed in some cases that the pollen commences its proper action by exserting its tube, and the germen commences swelling, though soon afterwards it decays. In the next stage in the series, hybrid offspring are produced though only rarely and few in number, and these are absolutely sterile: then we have hybrid offspring more numerous, and occasionally, though very rarely, breeding with either parent, as is the case with the common mule. Again, other hybrids, though infertile *inter se*, will breed *quite* freely with either parent, or with a third species, and will yield

[1] See Suchetet, *L'Hybridité dans la Nature*, Bruxelles, 1888, p. 67. In *Var. under Dom.*, Ed. ii. vol. II. hybrids between the fowl and the pheasant are mentioned. I can give no information on the other cases.

offspring generally infertile, but sometimes fertile; and these latter again will breed with either parent, or with a third or fourth species: thus Kölreuter blended together many forms. Lastly it is now admitted by those botanists who have longest contended against the admission, that in certain families the hybrid offspring of many of the species are sometimes perfectly fertile in the first generation when bred together: indeed in some few cases Mr Herbert[1] found that the hybrids were decidedly more fertile than either of their pure parents. There is no way to escape from the admission that the hybrids from some species of plants are fertile, except by declaring that no form shall be considered as a species, if it produces with another species fertile offspring: but this is begging the question[2]. It has often been stated that different species of animals have a sexual repugnance towards each other; I can find no evidence of this; it appears as if they merely did not excite each others passions. I do not believe that in this respect there is any essential distinction between animals and plants; and in the latter there cannot be a feeling of repugnance.

Causes of Sterility in Hybrids.

The difference in nature between species which causes the greater or lesser degree of sterility in their offspring appears, according to Herbert and Kölreuter, to be connected much less with external form, size, or structure, than with constitutional peculiarities; by which is meant their adaptation to different climates, food and situation, &c.: these

[1] *Origin*, Ed. i. p. 250, vi. p. 370.

[2] This was the position of Gärtner and of Kölreuter: see *Origin*, Ed. i. pp. 246–7, vi. pp. 367–8.

peculiarities of constitution probably affect the entire frame, and no one part in particular[1].

From the foregoing facts I think we must admit that there exists a perfect gradation in fertility between species which when crossed are quite fertile (as in Rhododendron, Calceolaria, &c.), and indeed in an extraordinary degree fertile (as in Crinum), and those species which never produce offspring, but which by certain effects (as the exsertion of the pollen-tube) evince their alliance. Hence, I conceive, we must give up sterility, although undoubtedly in a lesser or greater degree of very frequent occurrence, as an unfailing mark by which *species* can be distinguished from *races, i.e.* from those forms which have descended from a common stock.

Infertility from causes distinct from hybridisation.

Let us see whether there are any analogous facts which will throw any light on this subject, and will tend to explain why the offspring of certain species, when crossed, should be sterile, and not others, without requiring a distinct law connected with their creation to that effect. Great numbers, probably a large majority of animals when caught by man and removed from their natural conditions, although taken very young, rendered quite tame, living to a good old age, and apparently quite healthy, seem incapable under these circumstances of breeding[2]. I do not refer to animals kept in

[1] ⟨Note in the original.⟩ Yet this seems introductory to the case of the heaths and crocuses above mentioned. ⟨Herbert observed that crocus does not set seed if transplanted before pollination, but that such treatment after pollination has no sterilising effect. (*Var. under Dom.*, Ed. ii. vol. II. p. 148.) On the same page is a mention of the Ericaceæ being subject to contabescence of the anthers. For *Crinum* see *Origin*, Ed. i. p. 250: for *Rhododenron* and *Calceolaria* see p. 251.⟩

[2] ⟨Note in original.⟩ Animals seem more often made sterile by being

menageries, such as at the Zoological Gardens, many of which, however, appear healthy and live long and unite but do not produce; but to animals caught and left partly at liberty in their native country. Rengger[1] enumerates several caught young and rendered tame, which he kept in Paraguay, and which would not breed: the hunting leopard or cheetah and elephant offer other instances; as do bears in Europe, and the 25 species of hawks, belonging to different genera, thousands of which have been kept for hawking and have lived for long periods in perfect vigour. When the expense and trouble of procuring a succession of young animals in a wild state be borne in mind, one may feel sure that no trouble has been spared in endeavours to make them breed. So clearly marked is this difference in different kinds of animals, when captured by man, that St Hilaire makes two great classes of animals useful to man:—the *tame*, which will not breed, and the *domestic* which will breed in domestication. From certain singular facts we might have supposed that the non-breeding of animals was owing to some perversion of instinct. But we meet with exactly the same class of facts in plants: I do not refer to the large number of cases where the climate does not permit the seed or fruit to ripen, but where the flowers do not "set," owing to some imperfection of the ovule or pollen. The latter, which alone can be distinctly examined, is often manifestly imperfect, as any one with a microscope can observe by comparing the pollen of the Persian and Chinese lilacs[2] with the common lilac; the two

taken out of their native condition than plants, and so are more sterile when crossed.

We have one broad fact that sterility in hybrids is not closely related to external difference, and these are what man alone gets by selection.

[1] See *Var. under Dom.*, Ed. ii. vol. II. p. 132; for the case of the cheetah see *loc cit.* p. 133.

[2] *Var. under Dom.*, Ed. ii. vol. II. p. 148.

former species (I may add) are equally sterile in Italy as in this country. Many of the American bog plants here produce little or no pollen, whilst the Indian species of the same genera freely produce it. Lindley observes that sterility is the bane of the horticulturist[1]: Linnæus has remarked on the sterility of nearly all alpine flowers when cultivated in a lowland district[2]. Perhaps the immense class of double flowers chiefly owe their structure to an excess of food acting on parts rendered slightly sterile and less capable of performing their true function, and therefore liable to be rendered monstrous, which monstrosity, like any other disease, is inherited and rendered common. So far from domestication being in itself unfavourable to fertility, it is well known that when an organism is once capable of submission to such conditions ⟨its⟩ fertility is increased[3] beyond the natural limit. According to agriculturists, slight changes of conditions, that is of food or habitation, and likewise crosses with races slightly different, increase the vigour and probably the fertility of their offspring. It would appear also that even a great change of condition, for instance, transportal from temperate countries to India, in many cases does not in the least affect fertility, although it does health and length of life and the period of maturity. When sterility is induced by domestication it is of the same kind, and varies in degree, exactly as with hybrids: for be it remembered that the most sterile hybrid is no way monstrous; its organs are perfect, but they do not act, and minute microscopical investigations show that they are in the same state as those of pure species in the intervals of the breeding season. The defective pollen in the cases above alluded to pre-

1 Quoted in the *Origin*, Ed. i. p. 9.
2 See *Var. under Dom.*, Ed. ii. vol. II. p. 147.
3 *Var. under Dom.*, Ed. ii. vol. II. p. 89.

cisely resembles that of hybrids. The occasional breeding of hybrids, as of the common mule, may be aptly compared to the most rare but occasional reproduction of elephants in captivity. The cause of many exotic Geraniums producing (although in vigorous health) imperfect pollen seems to be connected with the period when water is given them[1]; but in the far greater majority of cases we cannot form any conjecture on what exact cause the sterility of organisms taken from their natural conditions depends. Why, for instance, the cheetah will not breed whilst the common cat and ferret (the latter generally kept shut up in a small box) do,—why the elephant will not whilst the pig will abundantly—why the partridge and grouse in their own country will not, whilst several species of pheasants, the guinea-fowl from the deserts of Africa and the peacock from the jungles of India, will. We must, however, feel convinced that it depends on some constitutional peculiarities in these beings not suited to their new condition; though not necessarily causing an ill state of health. Ought we then to wonder much that those hybrids which have been produced by the crossing of species with different constitutional tendencies (which tendencies we know to be eminently inheritable) should be sterile: it does not seem improbable that the cross from an alpine and lowland plant should have its constitutional powers deranged, in nearly the same manner as when the parent alpine plant is brought into a lowland district. Analogy, however, is a deceitful guide, and it would be rash to affirm, although it may appear probable, that the sterility of hybrids is due to the constitutional peculiarities of one parent being disturbed by being blended with those of the other parent in exactly the same

[1] See *Var. under Dom.*, Ed. ii. vol. II. p. 147.

manner as it is caused in some organic beings when placed by man out of their natural conditions[1]. Although this would be rash, it would, I think, be still rasher, seeing that sterility is no more incidental to *all* cross-bred productions than it is to all organic beings when captured by man, to assert that the sterility of certain hybrids proved a distinct creation of their parents.

But it may be objected[2] (however little the sterility of certain hybrids is connected with the distinct creations of species), how comes it, if species are only races produced by natural selection, that when crossed they so frequently produce sterile offspring, whereas in the offspring of those races confessedly produced by the arts of man there is no one instance of sterility. There is not much difficulty in this, for the races produced by the natural means above explained will be slowly but steadily selected; will be adapted to various and diverse conditions, and to these conditions they will be rigidly confined for immense periods of time; hence we may suppose that they would acquire different constitutional peculiarities adapted to the stations they occupy; and on the constitutional differences between species their sterility, according to the best authorities, depends. On the other hand man selects by external appearance[3]; from his ignorance, and from not having any test at least comparable in delicacy to the natural struggle for food, continued at intervals through the life of each individual, he cannot eliminate fine shades of constitution, dependent on invisible differences in the fluids or solids of the body; again, from the value

[1] *Origin*, Ed. i. p. 267, vi. p. 392. This is the principle experimentally investigated in the author's *Cross- and Self-Fertilisation*.

[2] *Origin*, Ed. i. p. 268, vi. p. 398.

[3] ⟨Notes in original.⟩ Mere difference of structure no guide to what will or will not cross. First step gained by races keeping apart. ⟨It is not clear where these notes were meant to go.⟩

which he attaches to each individual, he asserts his utmost power in contravening the natural tendency of the most vigorous to survive. Man, moreover, especially in the earlier ages, cannot have kept his conditions of life constant, and in later ages his stock pure. Until man selects two varieties from the same stock, adapted to two climates or to other different external conditions, and confines each rigidly for one or several thousand years to such conditions, always selecting the individuals best adapted to them, he cannot be said to have even commenced the experiment. Moreover, the organic beings which man has longest had under domestication have been those which were of the greatest use to him, and one chief element of their usefulness, especially in the earlier ages, must have been their capacity to undergo sudden transportals into various climates, and at the same time to retain their fertility, which in itself implies that in such respects their constitutional peculiarities were not closely limited. If the opinion already mentioned be correct, that most of the domestic animals in their present state have descended from the fertile commixture of wild races or species, we have indeed little reason now to expect infertility between any cross of stock thus descended.

It is worthy of remark, that as many organic beings, when taken by man out of their natural conditions, have their reproductive system ⟨so⟩ affected as to be incapable of propagation, so, we saw in the first chapter, that although organic beings when taken by man do propagate freely, their offspring after some generations vary or sport to a degree which can only be explained by their reproductive system being ⟨in⟩ some way affected. Again, when species cross, their offspring are generally sterile; but it was found by Kölreuter that when hybrids are capable of breeding with either parent, or with other species, that their offspring are subject after some generations to excessive variation[1]. Agriculturists, also, affirm that the offspring from mongrels, after the first generation, vary much. Hence we see that both sterility and variation in the succeeding generations are consequent both on the removal of individual species from their natural states and on species crossing. The connection between these facts may be accidental, but they certainly appear to elucidate and support each other,—on the principle of the reproductive system of all organic beings being eminently sensitive to any disturbance, whether from removal or commixture, in their constitutional relations to the conditions to which they are exposed.

Points of Resemblance between "Races" and "Species[2]*."*

Races and reputed species agree in some respects, although differing from causes which, we have seen, we can in some degree understand, in the fertility and "trueness" of their offspring. In the first place, there is no clear sign by which to distinguish races from species, as is evident from the great difficulty experienced by naturalists in attempting to discriminate them. As far as external characters are concerned, many of the races which are descended from the same stock differ far more than true species of the same genus; look at the willow-wrens, some of which skilful ornithologists can hardly distinguish from each other except by their nests; look at the wild swans, and compare the distinct species of these genera with the races of

[1] *Origin*, Ed. i. p. 272, vi. p. 404.

[2] This section seems not to correspond closely with any in the *Origin*, Ed. i.; in some points it resembles pp. 15, 16, also the section on analogous variation in distinct species, *Origin*, Ed. i. p. 159, vi. p. 194.

domestic ducks, poultry, and pigeons; and so again with plants, compare the cabbages, almonds, peaches and nectarines, &c. with the species of many genera. St Hilaire has even remarked that there is a greater difference in size between races, as in dogs (for he believes all have descended from one stock), than between the species of any one genus; nor is this surprising, considering that amount of food and consequently of growth is the element of change over which man has most power. I may refer to a former statement, that breeders believe the growth of one part or strong action of one function causes a decrease in other parts; for this seems in some degree analogous to the law of "organic compensation[1]," which many naturalists believe holds good. To give an instance of this law of compensation,—those species of Carnivora which have the canine teeth greatly developed have certain molar teeth deficient; or again, in that division of the Crustaceans in which the tail is much developed, the thorax is little so, and the converse. The points of difference between different races is often strikingly analogous to that between species of the same genus: trifling spots or marks of colour[2] (as the bars on pigeons' wings) are often preserved in races of plants and animals, precisely in the same manner as similar trifling characters often pervade all the species of a genus, and even of a family. Flowers in varying their colours often become veined and spotted and the leaves become divided like true species: it is known that the varieties of the same plant never have red, blue and yellow flowers, though the hyacinth makes a very near approach to an

[1] The law of compensation is discussed in the *Origin*, Ed. i. p. 147, vi. p. 182.

[2] ⟨Note in original.⟩ Boitard and Corbié on outer edging red in tail of bird,—so bars on wing, white or black or brown, or white edged with black or ⟨illegible⟩: analogous to marks running through genera but with different colours. Tail coloured in pigeons.

exception[1]; and different species of the same genus seldom, though sometimes they have flowers of these three colours. Dun-coloured horses having a dark stripe down their backs, and certain domestic asses having transverse bars on their legs, afford striking examples of a variation analogous in character to the distinctive marks of other species of the same genus.

External characters of Hybrids and Mongrels.

There is, however, as it appears to me, a more important method of comparison between species and races, namely the character of the offspring[2] when species are crossed and when races are crossed: I believe, in no one respect, except in sterility, is there any difference. It would, I think, be a marvellous fact, if species have been formed by distinct acts of creation, that they should act upon each other in uniting, like races descended from a common stock. In the first place, by repeated crossing one species can absorb and wholly obliterate the characters of another, or of several other species, in the same manner as one race will absorb by crossing another race. Marvellous, that one act of creation should absorb another or even several acts of creation! The offspring of species, that is hybrids, and the offspring of races, that is mongrels, resemble each other in being either intermediate in character (as is most frequent in hybrids) or in resembling sometimes closely one and sometimes the other parent; in both the offspring produced by the same act of conception sometimes differ in their

[1] ⟨Note in original.⟩ Oxalis and Gentian. ⟨In Gentians blue, yellow and reddish colours occur. In Oxalis yellow, purple, violet and pink.⟩

[2] This section corresponds roughly to that on *Hybrids and Mongrels compared independently of their fertility*, *Origin*, Ed. i. p. 272, vi. p. 403. The discussion on Gärtner's views, given in the *Origin*, is here wanting. The brief mention of prepotency is common to them both.

degree of resemblance; both hybrids and mongrels sometimes retain a certain part or organ very like that of either parent, both, as we have seen, become in succeeding generations variable; and this tendency to vary can be transmitted by both; in both for many generations there is a strong tendency to reversion to their ancestral form. In the case of a hybrid laburnum and of a supposed mongrel vine different parts of the same plants took after each of their two parents. In the hybrids from some species, and in the mongrel of some races, the offspring differ according as which of the two species, or of the two races, is the father (as in the common mule and hinny) and which the mother. Some races will breed together, which differ so greatly in size, that the dam often perishes in labour; so it is with some species when crossed; when the dam of one species has borne offspring to the male of another species, her succeeding offspring are sometimes stained (as in Lord Morton's mare by the quagga, wonderful as the fact[1] is) by this first cross; so agriculturists positively affirm is the case when a pig or sheep of one breed has produced offspring by the sire of another breed.

Summary of second chapter[2].

Let us sum up this second chapter. If slight variations do occur in organic beings in a state of nature; if changes of condition from geological causes do produce in the course of ages effects analogous to those of domestication on any, however few, organisms; and how can we doubt it,—from what is actually known, and from what may be presumed, since thousands of organisms taken by man

[1] See *Animals and Plants*, Ed. ii. vol. I. p. 435. The phenomenon of *Telegony*, supposed to be established by this and similar cases, is now generally discredited in consequence of Ewart's experiments.

[2] The section on p. 109 is an appendix to the summary

for sundry uses, and placed in new conditions, have varied. If such variations tend to be hereditary; and how can we doubt it,—when we see shades of expression, peculiar manners, monstrosities of the strangest kinds, diseases, and a multitude of other peculiarities, which characterise and form, being inherited, the endless races (there are 1200 kinds of cabbages[1]) of our domestic plants and animals. If we admit that every organism maintains its place by an almost periodically recurrent struggle; and how can we doubt it,—when we know that all beings tend to increase in a geometrical ratio (as is instantly seen when the conditions become for a time more favourable); whereas on an average the amount of food must remain constant, if so, there will be a natural means of selection, tending to preserve those individuals with any slight deviations of structure more favourable to the then existing conditions, and tending to destroy any with deviations of an opposite nature. If the above propositions be correct, and there be no law of nature limiting the possible amount of variation, new races of beings will,—perhaps only rarely, and only in some few districts,—be formed.

Limits of Variation.

That a limit to variation does exist in nature is assumed by most authors, though I am unable to discover a single fact on which this belief is grounded[2]. One of the commonest statements is that plants do not become acclimatised; and I have even observed that kinds not raised by seed, but propagated by cuttings, &c., are instanced. A good instance has, however, been advanced in the case of kidney beans, which it is believed are now as

[1] I do not know the authority for this statement.

[2] In the *Origin* no limit is placed to variation as far as I know.

tender as when first introduced. Even if we overlook the frequent introduction of seed from warmer countries, let me observe that as long as the seeds are gathered promiscuously from the bed, without continual observation and *careful* selection of those plants which have stood the climate best during their whole growth, the experiment of acclimatisation has hardly been begun. Are not all those plants and animals, of which we have the greatest number of races, the oldest domesticated? Considering the quite recent progress[1] of systematic agriculture and horticulture, is it not opposed to every fact, that we have exhausted the capacity of variation in our cattle and in our corn,—even if we have done so in some trivial points, as their fatness or kind of wool? Will any one say, that if horticulture continues to flourish during the next few centuries, that we shall not have numerous new kinds of the potato and Dahlia? But take two varieties of each of these plants, and adapt them to certain fixed conditions and prevent any cross for 5000 years, and then again vary their conditions; try many climates and situations; and who[2] will predict the number and degrees of difference which might arise from these stocks? I repeat that we know nothing of any limit to the possible amount of variation, and therefore to the number and differences of the races, which might be produced by the natural means of selection, so infinitely more efficient than the agency of man. Races thus produced would probably be very "true"; and if from having been adapted to different conditions of existence, they possessed different constitutions, if suddenly removed to some new station, they would perhaps be sterile and their offspring would perhaps be infertile.

[1] ⟨Note in original.⟩ History of pigeons shows increase of peculiarities during last years.

[2] Compare an obscure passage in the Essay of 1842, p. 14.

Such races would be undistinguishable from species. But is there any evidence that the species, which surround us on all sides, have been thus produced? This is a question which an examination of the economy of nature we might expect would answer either in the affirmative or negative[1].

[1] ⟨Note in original.⟩ Certainly ⟨two pages in the MS.⟩ ought to be here introduced, viz., difficulty in forming such organ, as eye, by selection. ⟨In the *Origin*, Ed. i., a chapter on *Difficulties on Theory* follows that on *Laws of Variation*, and precedes that on *Instinct*: this was also the arrangement in the Essay of 1842; whereas in the present Essay *Instinct* follows *Variation* and precedes *Difficulties*.⟩

CHAPTER III

ON THE VARIATION OF INSTINCTS AND OTHER MENTAL ATTRIBUTES UNDER DOMESTICATION AND IN STATE OF NATURE; ON THE DIFFICULTIES IN THIS SUBJECT; AND ON ANALOGOUS DIFFICULTIES WITH RESPECT TO CORPOREAL STRUCTURES

Variation of mental attributes under domestication.

I HAVE as yet only alluded to the mental qualities which differ greatly in different species. Let me here premise that, as will be seen in the Second Part, there is no evidence and consequently no attempt to show that *all* existing organisms have descended from any one common parent-stock, but that only those have so descended which, in the language of naturalists, are clearly related to each other. Hence the facts and reasoning advanced in this chapter do not apply to the first origin of the senses[1], or of the chief mental attributes, such as of memory, attention, reasoning, &c., &c., by which most or all of the great related groups are characterised, any more than they apply to the first origin of life, or growth, or the power of reproduction. The application of such facts as I have collected is merely to the differences of the primary mental qualities and of the instincts in the species[2] of the

[1] A similar proviso occurs in the chapter on instinct in *Origin*, Ed. i. p. 207, vi. p. 319.

[2] The discussion occurs later in Chapter VII of the *Origin*, Ed. i. than in the present Essay, where moreover it is fuller in some respects.

several great groups. In domestic animals every observer has remarked in how great a degree, in the individuals of the same species, the dispositions, namely courage, pertinacity, suspicion, restlessness, confidence, temper, pugnaciousness, affection, care of their young, sagacity, &c., &c., vary. It would require a most able metaphysician to explain how many primary qualities of the mind must be changed to cause these diversities of complex dispositions. From these dispositions being inherited, of which the testimony is unanimous, families and breeds arise, varying in these respects. I may instance the good and ill temper of different stocks of bees and of horses,—the pugnacity and courage of game fowls,—the pertinacity of certain dogs, as bull-dogs, and the sagacity of others,—for restlessness and suspicion compare a wild rabbit reared with the greatest care from its earliest age with the extreme tameness of the domestic breed of the same animal. The offspring of the domestic dogs which have run wild in Cuba[1], though caught quite young, are most difficult to tame, probably nearly as much so as the original parent-stock from which the domestic dog descended. The habitual "*periods*" of different families of the same species differ, for instance, in the time of year of reproduction, and the period of life when the capacity is acquired, and the hour of roosting (in Malay fowls), &c., &c. These periodical habits are perhaps essentially corporeal, and may be compared to nearly similar habits in plants, which are known to vary extremely. Consensual movements (as called by Müller) vary and are inherited,—such as the cantering and ambling paces in horses, the tumbling of pigeons, and perhaps the handwriting, which is sometimes so similar between father

[1] In the margin occurs the name of Poeppig. In *Var. under Dom.*, Ed. ii. vol. I. p. 28, the reference to Poeppig on the Cuban dogs contains no mention of the wildness of their offspring.

and sons, may be ranked in this class. *Manners*, and even tricks which perhaps are only *peculiar* manners, according to W. Hunter and my father, are distinctly inherited in cases where children have lost their parent in early infancy. The inheritance of expression, which often reveals the finest shades of character, is familiar to everyone.

Again the tastes and pleasures of different breeds vary, thus the shepherd-dog delights in chasing the sheep, but has no wish to kill them,—the terrier (see Knight) delights in killing vermin, and the spaniel in finding game. But it is impossible to separate their mental peculiarities in the way I have done: the tumbling of pigeons, which I have instanced as a consensual movement, might be called a trick and is associated with a taste for flying in a close flock at a great height. Certain breeds of fowls have a taste for roosting in trees. The different actions of pointers and setters might have been adduced in the same class, as might the peculiar *manner* of hunting of the spaniel. Even in the same breed of dogs, namely in fox-hounds, it is the fixed opinion of those best able to judge that the different pups are born with different tendencies; some are best to find their fox in the cover; some are apt to run straggling, some are best to make casts and to recover the lost scent, &c.; and that these peculiarities undoubtedly are transmitted to their progeny. Or again the tendency to point might be adduced as a distinct habit which has become inherited,—as might the tendency of a true sheep dog (as I have been assured is the case) to run round the flock instead of directly at them, as is the case with other young dogs when attempted to be taught. The "transandantes" sheep[1] in Spain, which for some centuries have been yearly taken a journey of several hundred miles from one province

[1] ⟨Note in original.⟩ Several authors.

to another, know when the time comes, and show the greatest restlessness (like migratory birds in confinement), and are prevented with difficulty from starting by themselves, which they sometimes do, and find their own way. There is a case on good evidence[1] of a sheep which, when she lambed, would return across a mountainous country to her own birth-place, although at other times of year not of a rambling disposition. Her lambs inherited this same disposition, and would go to produce their young on the farm whence their parent came; and so troublesome was this habit that the whole family was destroyed.

These facts must lead to the conviction, justly wonderful as it is, that almost infinitely numerous shades of disposition, of tastes, of peculiar movements, and even of individual actions, can be modified or acquired by one individual and transmitted to its offspring. One is forced to admit that mental phenomena (no doubt through their intimate connection with the brain) can be inherited, like infinitely numerous and fine differences of corporeal structure. In the same manner as peculiarities of corporeal structure slowly acquired or lost during mature life (especially cognisant ⟨?⟩ in disease), as well as congenital peculiarities, are transmitted; so it appears to be with the mind. The inherited paces in the horse have no doubt been acquired by compulsion during the lives of the parents: and temper and tameness may be modified in a breed by the treatment which the individuals receive. Knowing that a pig has been taught to point, one would suppose that this quality in pointer-dogs was the simple result of habit, but some facts, with respect to the occasional appearance of a similar quality in other dogs, would make one suspect that it originally

[1] In the margin "Hogg" occurs as authority for this fact. For the reference, see p. 17, note 4.

appeared in a less perfect degree, "*by chance*," that is from a congenital tendency[1] in the parent of the breed of pointers. One cannot believe that the tumbling, and high flight in a compact body, of one breed of pigeons has been taught; and in the case of the slight differences in the manner of hunting in young fox-hounds, they are doubtless congenital. The inheritance of the foregoing and similar mental phenomena ought perhaps to create less surprise, from the reflection that in no case do individual acts of reasoning, or movements, or other phenomena connected with consciousness, appear to be transmitted. An action, even a very complicated one, when from long practice it is performed unconsciously without any effort (and indeed in the case of many peculiarities of manners opposed to the will) is said, according to a common expression, to be performed "instinctively." Those cases of languages, and of songs, learnt in early childhood and *quite* forgotten, being *perfectly* repeated during the unconsciousness of illness, appear to me only a few degrees less wonderful than if they had been transmitted to a second generation[2].

Hereditary habits compared with instincts.

The chief characteristics of true instincts appear to be their invariability and non-improvement during the mature age of the individual animal: the absence of knowledge of the end, for which the action is performed, being associated, however, sometimes with a degree of reason; being subject to mistakes and

[1] In the *Origin*, Ed. i., he speaks more decidedly against the belief that instincts are hereditary habits, see for instance pp. 209, 214, Ed. vi. pp. 321, 327. He allows, however, something to habit (p. 216).

[2] A suggestion of Hering's and S. Butler's views on memory and inheritance. It is not, however, implied that Darwin was inclined to accept these opinions.

being associated with certain states of the body or times of the year or day. In most of these respects there is a resemblance in the above detailed cases of the mental qualities acquired or modified during domestication. No doubt the instincts of wild animals are more uniform than those habits or qualities modified or recently acquired under domestication, in the same manner and from the same causes that the corporeal structure in this state is less uniform than in beings in their natural conditions. I have seen a young pointer point as fixedly, the first day it was taken out, as any old dog; Magendie says this was the case with a retriever which he himself reared: the tumbling of pigeons is not probably improved by age: we have seen that in the case above given that the young sheep inherited the migratory tendency to their particular birth-place the first time they lambed. This last fact offers an instance of a domestic instinct being associated with a state of body; as do the "transandantes" sheep with a time of year. Ordinarily the acquired instincts of domestic animals seem to require a certain degree of education (as generally in pointers and retrievers) to be perfectly developed: perhaps this holds good amongst wild animals in rather a greater degree than is generally supposed; for instance, in the singing of birds, and in the knowledge of proper herbs in Ruminants. It seems pretty clear that bees transmit knowledge from generation to generation. Lord Brougham[1] insists strongly on ignorance of the end proposed being eminently characteristic of true instincts; and this appears to me to apply to many acquired hereditary habits; for instance, in the case of the young pointer alluded to before, which pointed so steadfastly the first day that we were obliged several times to carry

[1] Lord Brougham's *Dissertations on Subjects of Science*, etc., 1839, p. 27.

him away[1]. This puppy not only pointed at sheep, at large white stones, and at every little bird, but likewise "backed" the other pointers: this young dog must have been as unconscious for what end he was pointing, namely to facilitate his master's killing game to eat, as is a butterfly which lays her eggs on a cabbage, that her caterpillars would eat the leaves. So a horse that ambles instinctively, manifestly is ignorant that he performs that peculiar pace for the ease of man; and if man had never existed, he would never have ambled. The young pointer pointing at white stones appears to be as much a mistake of its acquired instinct, as in the case of flesh-flies laying their eggs on certain flowers instead of putrifying meat. However true the ignorance of the end may generally be, one sees that instincts are associated with some degree of reason; for instance, in the case of the tailor-bird, who spins threads with which to make her nest ⟨yet⟩ will use artificial threads when she can procure them[2]; so it has been known that an old pointer has broken his point and gone round a hedge to drive out a bird towards his master[3].

There is one other quite distinct method by which the instincts or habits acquired under domestication may be compared with those given by nature, by a test of a fundamental kind; I mean the comparison of the mental powers of mongrels and hybrids. Now the instincts, or habits, tastes, and dispositions of one *breed* of animals, when crossed with another breed, for instance a shepherd-

[1] This case is more briefly given in the *Origin*, Ed. i. p. 213, vi. p. 326. The simile of the butterfly occurs there also.

[2] "A little dose, as Pierre Huber expresses it, of judgment or reason, often comes into play." *Origin*, Ed. i. p. 208, vi. p. 320.

[3] In the margin is written "Retriever killing one bird." This refers to the cases given in the *Descent of Man*, 2nd Ed. (in 1 vol.) p. 78, of a retriever being puzzled how to deal with a wounded and a dead bird, killed the former and carried both at once. This was the only known instance of her wilfully injuring game.

dog with a harrier, are blended and appear in the same curiously mixed degree, both in the first and succeeding generations, exactly as happens when one *species* is crossed with another[1]. This would hardly be the case if there was any fundamental difference between the domestic and natural instinct[2]; if the former were, to use a metaphorical expression, merely superficial.

Variation in the mental attributes of wild animals.

With respect to the variation[3] of the mental powers of animals in a wild state, we know that there is a considerable difference in the disposition of different individuals of the same species, as is recognised by all those who have had the charge of animals in a menagerie. With respect to the wildness of animals, that is fear directed particularly against man, which appears to be as true an instinct as the dread of a young mouse of a cat, we have excellent evidence that it is slowly acquired and becomes hereditary. It is also certain that, in a natural state, individuals of the same species lose

[1] See *Origin*, Ed. i. p. 214, vi. p. 327.

[2] ⟨Note in original.⟩ Give some definition of instinct, or at least give chief attributes. ⟨In *Origin*, Ed. i. p. 207, vi. p. 319, Darwin refuses to define instinct.⟩ The term instinct is often used in ⟨a⟩ sense which implies no more than that the animal does the action in question. Faculties and instincts may I think be imperfectly separated. The mole has the faculty of scratching burrows, and the instinct to apply it. The bird of passage has the faculty of finding its way and the instinct to put it in action at certain periods. It can hardly be said to have the faculty of knowing the time, for it can possess no means, without indeed it be some consciousness of passing sensations. Think over all habitual actions and see whether faculties and instincts can be separated. We have faculty of waking in the night, if an instinct prompted us to do something at certain hour of night or day. Savages finding their way. Wrangel's account—probably a faculty inexplicable by the possessor. There are besides faculties "*means,*" as conversion of larvæ into neuters and queens. I think all this generally implied, anyhow useful. ⟨This discussion, which does not occur in the *Origin*, is a first draft of that which follows in the text, p. 123.⟩

[3] A short discussion of a similar kind occurs in the *Origin*, Ed. i. p. 211, vi. p. 324.

or do not practice their migratory instincts—as woodcocks in Madeira. With respect to any variation in the more complicated instincts, it is obviously most difficult to detect, even more so than in the case of corporeal structure, of which it has been admitted the variation is exceedingly small, and perhaps scarcely any in the majority of species at any one period. Yet, to take one excellent case of instinct, namely the nests of birds, those who have paid most attention to the subject maintain that not only certain individuals (? species) seem to be able to build very imperfectly, but that a difference in skill may not unfrequently be detected between individuals[1]. Certain birds, moreover, adapt their nests to circumstances; the water-ouzel makes no vault when she builds under cover of a rock—the sparrow builds very differently when its nest is in a tree or in a hole, and the golden-crested wren sometimes suspends its nest below and sometimes places it *on* the branches of trees.

Principles of Selection applicable to instincts.

As the instincts of a species are fully as important to its preservation and multiplication as its corporeal structure, it is evident that if there be the slightest congenital differences in the instincts and habits, or if certain individuals during their lives are induced or compelled to vary their habits, and if such differences are in the smallest degree more favourable, under slightly modified external conditions, to their preservation, such individuals must in the long run have a better *chance* of being preserved and of multiplying[2]. If this be admitted, a series of small changes may, as in the case of corporeal structure, work great changes in the mental powers, habits and instincts of any species.

[1] This sentence agrees with the MS., but is clearly in need of correction.
[2] This corresponds to *Origin*, Ed. i. p. 212, vi. p. 325.

Difficulties in the acquirement of complex instincts by Selection.

Every one will at first be inclined to explain (as I did for a long time) that many of the more complicated and wonderful instincts could not be acquired in the manner here supposed[1]. The Second Part of this work is devoted to the general consideration of how far the general economy of nature justifies or opposes the belief that related species and genera are descended from common stocks; but we may here consider whether the instincts of animals offer such a *primâ facie* case of impossibility of gradual acquirement, as to justify the rejection of any such theory, however strongly it may be supported by other facts. I beg to repeat that I wish here to consider not the *probability* but the *possibility* of complicated instincts having been acquired by the slow and long-continued selection of very slight (either congenital or produced by habit) modifications of foregoing simpler instincts; each modification being as useful and necessary, to the species practising it, as the most complicated kind.

First, to take the case of birds'-nests; of existing species (almost infinitely few in comparison with the multitude which must have existed, since the period of the new Red Sandstone of N. America, of whose habits we must always remain ignorant) a tolerably perfect series could be made from eggs

[1] This discussion is interesting in differing from the corresponding section of the *Origin*, Ed. i. p. 216, vi. p. 330, to the end of the chapter. In the present Essay the subjects dealt with are nest-making instincts, including the egg-hatching habit of the Australian bush-turkey. The power of "shamming death." "Faculty" in relation to instinct. The instinct of lapse of time, and of direction. Bees' cells very briefly given. Birds feeding their young on food differing from their own natural food. In the *Origin*, Ed. i., the cases discussed are the instinct of laying eggs in other birds' nests; the slave-making instinct in ants; the construction of the bee's comb, very fully discussed.

laid on the bare ground, to others with a few sticks just laid round them, to a simple nest like the wood-pigeons, to others more and more complicated: now if, as is asserted, there occasionally exist slight differences in the building powers of an individual, and if, which is at least probable, that such differences would tend to be inherited, then we can see that it is at least *possible* that the nidificatory instincts may have been acquired by the gradual selection, during thousands and thousands of generations, of the eggs and young of those individuals, whose nests were in some degree better adapted to the preservation of their young, under the then existing conditions. One of the most surprising instincts on record is that of the Australian bush-turkey, whose eggs are hatched by the heat generated from a huge pile of fermenting materials, which it heaps together; but here the habits of an allied species show how this instinct *might possibly* have been acquired. This second species inhabits a tropical district, where the heat of the sun is sufficient to hatch its eggs; this bird, burying its eggs, apparently for concealment, under a lesser heap of rubbish, but of a dry nature, so as not to ferment. Now suppose this bird to range slowly into a climate which was cooler, and where leaves were more abundant, in that case, those individuals, which chanced to have their collecting instinct strongest developed, would make a somewhat larger pile, and the eggs, aided during some colder season, under the slightly cooler climate by the heat of incipient fermentation, would in the long run be more freely hatched and would probably produce young ones with the same more highly developed collecting tendencies; of these again, those with the best developed powers would again tend to rear most young. Thus this strange instinct might *possibly* be acquired, every individual bird being as ignorant of the laws of fermentation, and the consequent development of heat, as we know they must be.

Secondly, to take the case of animals feigning death (as it is commonly expressed) to escape danger. In the case of insects, a perfect series can be shown, from some insects, which momentarily stand still, to others which for a second slightly contract their legs, to others which will remain immovably drawn together for a quarter of an hour, and may be torn asunder or roasted at a slow fire, without evincing the smallest sign of sensation. No one will doubt that the length of time, during which each remains immovable, is well adapted to ⟨favour the insect's⟩ escape ⟨from⟩ the dangers to which it is most exposed, and few will deny the *possibility* of the change from one degree to another, by the means and at the rate already explained. Thinking it, however, wonderful (though not impossible) that the attitude of death should have been acquired by methods which imply no imitation, I compared several species, when feigning, as is said, death, with others of the same species really dead, and their attitudes were in no one case the same.

Thirdly, in considering many instincts it is useful to *endeavour* to separate the faculty[1] by which they perform it, and the mental power which urges to the performance, which is more properly called an instinct. We have an instinct to eat, we have jaws &c. to give us the faculty to do so. These faculties are often unknown to us: bats, with their eyes destroyed, can avoid strings suspended across a room, we know not at present by what faculty they do this. Thus also, with migratory birds, it is a

[1] The distinction between *faculty* and *instinct* corresponds in some degree to that between perception of a stimulus and a specific reaction. I imagine that the author would have said that the sensitiveness to light possessed by a plant is *faculty*, while *instinct* decides whether the plant curves to or from the source of illumination.

wonderful instinct which urges them at certain times of the year to direct their course in certain directions, but it is a faculty by which they know the time and find their way. With respect to time[1], man without seeing the sun can judge to a certain extent of the hour, as must those cattle which come down from the inland mountains to feed on sea-weed left bare at the changing hour of low-water[2]. A hawk (D'Orbigny) seems certainly to have acquired a knowledge of a period of every 21 days. In the cases already given of the sheep which travelled to their birth-place to cast their lambs, and the sheep in Spain which know their time of march[3], we may conjecture that the tendency to move is associated, we may then call it instinctively, with some corporeal sensations. With respect to direction we can easily conceive how a tendency to travel in a certain course may possibly have been acquired, although we must remain ignorant how birds are able to preserve any direction whatever in a dark night over the wide ocean. I may observe that the power of some savage races of mankind to find their way, although perhaps wholly different from the faculty of birds, is nearly as unintelligible to us. Bellinghausen, a skilful navigator, describes with the utmost wonder the manner in which some Esquimaux guided him to a certain point, by a course never straight, through newly formed hummocks of ice, on a thick foggy day, when he with a compass found it impossible, from having no landmarks, and from their course being so extremely crooked, to preserve any sort of uniform

[1] ⟨Note in the original in an unknown handwriting.⟩ At the time when corn was pitched in the market instead of sold by sample, the geese in the town fields of Newcastle ⟨Staffordshire?⟩ used to know market day and come in to pick up the corn spilt.

[2] ⟨Note in original.⟩ Macculloch and others.

[3] I can find no reference to the *transandantes* sheep in Darwin's published work. He was possibly led to doubt the accuracy of the statement on which he relied. For the case of the sheep returning to their birth-place see p. 17, note 4.

direction: so it is with Australian savages in thick forests. In North and South America many birds slowly travel northward and southward, urged on by the food they find, as the seasons change; let them continue to do this, till, as in the case of the sheep in Spain, it has become an urgent instinctive desire, and they will gradually accelerate their journey. They would cross narrow rivers, and if these were converted by subsidence into narrow estuaries, and gradually during centuries to arms of the sea, still we may suppose their restless desire of travelling onwards would impel them to cross such an arm, even if it had become of great width beyond their span of vision. How they are able to preserve a course in any direction, I have said, is a faculty unknown to us. To give another illustration of the means by which I conceive it *possible* that the direction of migrations have been determined. Elk and reindeer in N. America annually cross, as if they could marvellously smell or see at the distance of a hundred miles, a wide tract of absolute desert, to arrive at certain islands where there is a scanty supply of food; the changes of temperature, which geology proclaims, render it probable that this desert tract formerly supported some vegetation, and thus these quadrupeds might have been annually led on, till they reached the more fertile spots, and so acquired, like the sheep of Spain, their migratory powers.

Fourthly, with respect to the combs of the hive-bee[1]; here again we must look to some faculty or means by which they make their hexagonal cells, without indeed we view these instincts as mere machines. At present such a faculty is quite unknown: Mr Waterhouse supposes that several bees are led by their instinct to excavate a mass of wax to a certain thinness, and that the result of this

[1] *Origin,* Ed. i. p. 224, vi. p. 342.

is that hexagons necessarily remain. Whether this or some other theory be true, some such means they must possess. They abound, however, with true instincts, which are the most wonderful that are known. If we examine the little that is known concerning the habits of other species of bees, we find much simpler instincts: the humble bee merely fills rude balls of wax with honey and aggregates them together with little order in a rough nest of grass. If we knew the instinct of all the bees, which ever had existed, it is not improbable that we should have instincts of every degree of complexity, from actions as simple as a bird making a nest, and rearing her young, to the wonderful architecture and government of the hive-bee; at least such is *possible*, which is all that I am here considering.

Finally, I will briefly consider under the same point of view one other class of instincts, which have often been advanced as truly wonderful, namely parents bringing food to their young which they themselves neither like nor partake of[1];—for instance, the common sparrow, a granivorous bird, feeding its young with caterpillars. We might of course look into the case still earlier, and seek how an instinct in the parent, of feeding its young at all, was first derived; but it is useless to waste time in conjectures on a series of gradations from the young feeding themselves and being slightly and occasionally assisted in their search, to their entire food being brought to them. With respect to the parent bringing a different kind of food from its own kind, we may suppose either that the remote stock, whence the sparrow and other congenerous birds have descended, was insectivorous, and that its own habits and structure have been changed, whilst its ancient instincts with respect to its young have remained

[1] This is an expansion of an obscure passage in the Essay of 1842, p. 19.

unchanged; or we may suppose that the parents have been induced to vary slightly the food of their young, by a slight scarcity of the proper kind (or by the instincts of some individuals not being so truly developed), and in this case those young which were most capable of surviving were necessarily most often preserved, and would themselves in time become parents, and would be similarly compelled to alter their food for their young. In the case of those animals, the young of which feed themselves, changes in their instincts for food, and in their structure, might be selected from slight variations, just as in mature animals. Again, where the food of the young depends on where the mother places her eggs, as in the case of the caterpillars of the cabbage-butterfly, we may suppose that the parent stock of the species deposited her eggs sometimes on one kind and sometimes on another of congenerous plants (as some species now do), and if the cabbage suited the caterpillars better than any other plant, the caterpillars of those butterflies, which had chosen the cabbage, would be most plentifully reared, and would produce butterflies more apt to lay their eggs on the cabbage than on the other congenerous plants.

However vague and unphilosophical these conjectures may appear, they serve, I think, to show that one's first impulse utterly to reject any theory whatever, implying a gradual acquirement of these instincts, which for ages have excited man's admiration, may at least be delayed. Once grant that dispositions, tastes, actions or habits can be slightly modified, either by slight congenital differences (we must suppose in the brain) or by the force of external circumstances, and that such slight modifications can be rendered inheritable,—a proposition which no one can reject,—and it will be difficult to put any limit to the complexity and wonder of the tastes and habits which may *possibly* be thus acquired.

Difficulties in the acquirement by Selection of complex corporeal structures.

After the past discussion it will perhaps be convenient here to consider whether any particular corporeal organs, or the entire structure of any animals, are so wonderful as to justify the rejection *primâ facie* of our theory[1]. In the case of the eye, as with the more complicated instincts, no doubt one's first impulse is to utterly reject every such theory. But if the eye from its most complicated form can be shown to graduate into an exceedingly simple state,—if selection can produce the smallest change, and if such a series exists, then it is clear (for in this work we have nothing to do with the first origin of organs in their simplest forms[2]) that it may *possibly* have been acquired by gradual selection of slight, but in each case, useful deviations[3]. Every naturalist, when he meets with any new and singular organ, always expects to find, and looks for, other and simpler modifications of it in other beings. In the case of the eye, we have a multitude of different forms, more or less simple, not graduating

[1] The difficulties discussed in the *Origin*, Ed. i. p. 171, vi. p. 207, are the rarity of transitional varieties, the origin of the tail of the giraffe; the otter-like polecat (*Mustela vison*); the flying habit of the bat; the penguin and the logger-headed duck; flying fish; the whale-like habit of the bear; the woodpecker; diving petrels; the eye; the swimming bladder; Cirripedes; neuter insects; electric organs.

Of these, the polecat, the bat, the woodpecker, the eye, the swimming bladder are discussed in the present Essay, and in addition some botanical problems.

[2] In the *Origin*, Ed. vi. p. 275, the author replies to Mivart's criticisms (*Genesis of Species*, 1871), referring especially to that writer's objection "that natural selection is incompetent to account for the incipient stages of useful structures."

[3] ⟨The following sentence seems to have been intended for insertion here⟩ "and that each eye throughout the animal kingdom is not only most useful, but *perfect* for its possessor."

into each other, but separated by sudden gaps or intervals; but we must recollect how incomparably greater would the multitude of visual structures be if we had the eyes of every fossil which ever existed. We shall discuss the probable vast proportion of the extinct to the recent in the succeeding Part. Notwithstanding the large series of existing forms, it is most difficult even to conjecture by what intermediate stages very many simple organs could possibly have graduated into complex ones: but it should be here borne in mind, that a part having originally a wholly different function, may on the theory of gradual selection be slowly worked into quite another use; the gradations of forms, from which naturalists believe in the hypothetical metamorphosis of part of the ear into the swimming bladder in fishes[1], and in insects of legs into jaws, show the manner in which this is possible. As under domestication, modifications of structure take place, without any continued selection, which man finds very useful, or valuable for curiosity (as the hooked calyx of the teazle, or the ruff round some pigeons' necks), so in a state of nature some small modifications, apparently beautifully adapted to certain ends, may perhaps be produced from the accidents of the reproductive system, and be at once propagated without long-continued selection of small deviations towards that structure[2]. In conjecturing by what stages any complicated organ in a species may have arrived at its present state, although we may look to the analogous organs in other existing species, we should do this merely to aid and guide our imaginations; for to know the real stages we

[1] *Origin*, Ed. i. p. 190, vi. p. 230.

[2] This is one of the most definite statements in the present Essay of the possible importance of *sports* or what would now be called *mutations*. As is well known the author afterwards doubted whether species could arise in this way. See *Origin*, Ed. v. p. 103, vi. p. 110, also *Life and Letters*, vol. iii. p. 107.

must look only through one line of species, to one ancient stock, from which the species in question has descended. In considering the eye of a quadruped, for instance, though we may look at the eye of a molluscous animal or of an insect, as a proof how simple an organ will serve some of the ends of vision; and at the eye of a fish as a nearer guide of the manner of simplication; we must remember that it is a mere chance (assuming for a moment the truth of our theory) if any existing organic being has preserved any one organ, in exactly the same condition, as it existed in the ancient species at remote geological periods.

The nature or condition of certain structures has been thought by some naturalists to be of no use to the possessor[1], but to have been formed wholly for the good of other species; thus certain fruit and seeds have been thought to have been made nutritious for certain animals—numbers of insects, especially in their larval state, to exist for the same end—certain fish to be bright coloured to aid certain birds of prey in catching them, &c. Now could this be proved (which I am far from admitting) the theory of natural selection would be quite overthrown; for it is evident that selection depending on the advantage over others of one individual with some slight deviation would never produce a structure or quality profitable only to another species. No doubt one being takes advantage of qualities in another, and may even cause its extermination; but this is far from proving that this quality was produced for such an end. It may be advantageous to a plant to have its seeds attractive to animals, if one out of a hundred or a thousand escapes being

[1] See *Origin*, Ed. i. p. 210, vi. p. 322, where the question is discussed for the case of instincts with a proviso that the same argument applies to structure. It is briefly stated in its general bearing in *Origin*, Ed. i. p. 87, vi. p. 106.

digested, and thus aids dissemination: the bright colours of a fish may be of some advantage to it, or more probably may result from exposure to certain conditions in favourable haunts for food, *notwithstanding* it becomes subject to be caught more easily by certain birds.

If instead of looking, as above, at certain individual organs, in order to speculate on the stages by which their parts have been matured and selected, we consider an individual animal, we meet with the same or greater difficulty, but which, I believe, as in the case of single organs, rests entirely on our ignorance. It may be asked by what intermediate forms could, for instance, a bat possibly have passed; but the same question might have been asked with respect to the seal, if we had not been familiar with the otter and other semi-aquatic carnivorous quadrupeds. But in the case of the bat, who can say what might have been the habits of some parent form with less developed wings, when we now have insectivorous opossums and herbivorous squirrels fitted for merely gliding through the air[1]. One species of bat is at present partly aquatic in its habits[2]. Woodpeckers and tree-frogs are especially adapted, as their names express, for climbing trees; yet we have species of both inhabiting the open plains of La Plata, where a tree does not exist[3]. I might argue from this circumstance that a structure eminently fitted for climbing trees might descend from forms inhabiting a country where a tree

[1] ⟨Note in original.⟩ No one will dispute that the gliding is most useful, probably necessary for the species in question.

[2] ⟨Note in original.⟩ Is this the Galeopithecus? I forget. ⟨*Galeopithecus* "or the flying Lemur" is mentioned in the corresponding discussion in the *Origin*, Ed. i. p. 181, vi. p. 217, as formerly placed among the bats. I do not know why it is described as partly aquatic in its habits.⟩

[3] In the *Origin*, Ed. vi. p. 221, the author modified the statement that it *never* climbs trees; he also inserted a sentence quoting Mr Hudson to the effect that in other districts this woodpecker climbs trees and bores holes. See Mr Darwin's paper, *Zoolog. Soc. Proc.*, 1870, and *Life and Letters*, iii. p. 153.

did not exist. Notwithstanding these and a multitude of other well-known facts, it has been maintained by several authors that one species, for instance of the carnivorous order, could not pass into another, for instance into an otter, because in its transitional state its habits would not be adapted to any proper conditions of life; but the jaguar[1] is a thoroughly terrestrial quadruped in its structure, yet it takes freely to the water and catches many fish; will it be said that it is *impossible* that the conditions of its country might become such that the jaguar should be driven to feed more on fish than they now do; and in that case is it impossible, is it not probable, that any the slightest deviation in its instincts, its form of body, in the width of its feet, and in the extension of the skin (which already unites the base of its toes) would give such individuals a better *chance* of surviving and propagating young with similar, barely perceptible (though thoroughly exercised), deviations[2]? Who will say what could thus be effected in the course of ten thousand generations? Who can answer the same question with respect to instincts? If no one can, the *possibility* (for we are not in this chapter considering the *probability*) of simple organs or organic beings being modified by natural selection and the effects of external agencies into complicated ones ought not to be absolutely rejected.

[1] Note by the late Alfred Newton. Richardson in *Fauna Boreali-Americana*, i. p. 49.

[2] ⟨Note in original.⟩ See Richardson a far better case of a polecat animal ⟨*Mustela vison*⟩, which half-year is aquatic. ⟨Mentioned in *Origin*, Ed. i. p. 179, vi. p. 216.⟩

PART II[1]

ON THE EVIDENCE FAVOURABLE AND OPPOSED TO THE VIEW THAT SPECIES ARE NATURALLY FORMED RACES, DESCENDED FROM COMMON STOCKS

CHAPTER IV

ON THE NUMBER OF INTERMEDIATE FORMS REQUIRED ON THE THEORY OF COMMON DESCENT; AND ON THEIR ABSENCE IN A FOSSIL STATE

I MUST here premise that, according to the view ordinarily received, the myriads of organisms, which have during past and present times peopled this world, have been created by so many distinct acts of creation. It is impossible to reason concerning the will of the Creator, and therefore, according to this view, we can see no cause why or why not the individual organism should have been created on any fixed scheme. That all the organisms of this world have been produced on a scheme is certain from their general affinities; and if this scheme can be shown to be the same with that which would result from allied organic beings descending from common stocks, it becomes highly improbable that they have been separately created by individual acts of the will of a Creator. For as well might it be said that, although the planets move in courses conformably to the law of gravity, yet we ought to

[1] In the *Origin* the division of the work into Parts I and II is omitted. In the MS. the chapters of Part II are numbered afresh, the present being Ch. I of Pt. II. I have thought it best to call it Ch. IV and there is evidence that Darwin had some thought of doing the same. It corresponds to Ch. IX of *Origin*, Ed. i., Ch. X in Ed. vi.

attribute the course of each planet to the individual act of the will of the Creator[1]. It is in every case more conformable with what we know of the government of this earth, that the Creator should have imposed only general laws. As long as no method was known by which races could become exquisitely adapted to various ends, whilst the existence of species was thought to be proved by the sterility[2] of their offspring, it was allowable to attribute each organism to an individual act of creation. But in the two former chapters it has (I think) been shown that the production, under existing conditions, of exquisitely adapted species, is at least *possible*. Is there then any direct evidence in favour ⟨of⟩ or against this view? I believe that the geographical distribution of organic beings in past and present times, the kind of affinity linking them together, their so-called "metamorphic" and "abortive" organs, appear in favour of this view. On the other hand, the imperfect evidence of the continuousness of the organic series, which, we shall immediately see, is required on our theory, is against it; and is the most weighty objection[3]. The evidence, however, even on this point, as far as it goes, is favourable; and considering the imperfection of our knowledge, especially with respect to past ages, it would be surprising if evidence drawn from such sources were not also imperfect.

As I suppose that species have been formed in

[1] In the Essay of 1842 the author uses astronomy in the same manner as an illustration. In the *Origin* this does not occur; the reference to the action of secondary causes is more general, *e.g.* Ed. i. p. 488, vi. p. 668.

[2] It is interesting to find the argument from sterility given so prominent a place. In a corresponding passage in the *Origin*, Ed. i. p. 480, vi. p. 659, it is more summarily treated. The author gives, as the chief bar to the acceptance of evolution, the fact that "we are always slow in admitting any great change of which we do not see the intermediate steps"; and goes on to quote Lyell on geological action. It will be remembered that the question of sterility remained a difficulty for Huxley.

[3] Similar statements occur in the Essay of 1842, p. 24, note 1, and in the *Origin*, Ed. i. p. 299.

an analogous manner with the varieties of the domesticated animals and plants, so must there have existed intermediate forms between all the species of the same group, not differing more than recognised varieties differ. It must not be supposed necessary that there should have existed forms exactly intermediate in character between any two species of a genus, or even between any two varieties of a species; but it is necessary that there should have existed every intermediate form between the one species or variety of the common parent, and likewise between the second species or variety, and this same common parent. Thus it does not necessarily follow that there ever has existed ⟨a⟩ series of intermediate sub-varieties (differing no more than the occasional seedlings from the same seed-capsule, between broccoli and common red cabbage; but it is certain that there has existed, between broccoli and the wild parent cabbage, a series of such intermediate seedlings, and again between red cabbage and the wild parent cabbage: so that the broccoli and red cabbage are linked together, but not *necessarily* by directly intermediate forms[1]. It is of course possible that there *may* have been directly intermediate forms, for the broccoli may have long since descended from a common red cabbage, and this from the wild cabbage. So on my theory, it must have been with species of the same genus. Still more must the supposition be avoided that there has necessarily ever existed (though one *may* have descended from ⟨the⟩ other) directly intermediate forms between any two genera or families—for instance between the genus *Sus* and the Tapir[2]; although it is necessary that intermediate forms (not differing more than the varieties

[1] In the *Origin*, Ed. i. p. 280, vi. p. 414 he uses his newly-acquired knowledge of pigeons to illustrate this point.

[2] Compare the *Origin*, Ed. i. p. 281, vi. p. 414.

of our domestic animals) should have existed between Sus and some unknown parent form, and Tapir with this same parent form. The latter may have differed more from Sus and Tapir than these two genera now differ from each other. In this sense, according to our theory, there has been a gradual passage (the steps not being wider apart than our domestic varieties) between the species of the same genus, between genera of the same family, and between families of the same order, and so on, as far as facts, hereafter to be given, lead us; and the number of forms which must have at former periods existed, thus to make good this passage between different species, genera, and families, must have been almost infinitely great.

What evidence[1] is there of a number of intermediate forms having existed, making a passage in the above sense, between the species of the same groups? Some naturalists have supposed that if every fossil which now lies entombed, together with all existing species, were collected together, a perfect series in every great class would be formed. Considering the enormous number of species requisite to effect this, especially in the above sense of the forms not being *directly* intermediate between the existing species and genera, but only intermediate by being linked through a common but often widely different ancestor, I think this supposition highly improbable. I am however far from underrating the probable number of fossilised species: no one who has attended to the wonderful progress of palæontology during the last few years will doubt that we as yet have found only an exceedingly small fraction of the species buried in the crust of the earth. Although the almost infinitely numerous intermediate forms in no one

[1] *Origin*, Ed. i. p. 301, vi. p. 440.

class may have been preserved, it does not follow that they have not existed. The fossils which have been discovered, it is important to remark, do tend, the little way they go, to make good the series; for as observed by Buckland they all fall into or between existing groups[1]. Moreover, those that fall between our existing groups, fall in, according to the manner required by our theory, for they do not directly connect two existing species of different groups, but they connect the groups themselves: thus the Pachydermata and Ruminantia are now separated by several characters, ⟨for instance⟩ the Pachydermata[2] have both a tibia and fibula, whilst Ruminantia have only a tibia; now the fossil Macrauchenia has a leg bone exactly intermediate in this respect, and likewise has some other intermediate characters. But the Macrauchenia does not connect any one species of Pachydermata with some one other of Ruminantia but it shows that these two groups have at one time been less widely divided. So have fish and reptiles been at one time more closely connected in some points than they now are. Generally in those groups in which there has been most change, the more ancient the fossil, if not identical with recent, the more often it falls between existing groups, or into small existing groups which now lie between other large existing groups. Cases like the foregoing, of which there are many, form steps, though few and far between, in a series of the kind required by my theory.

As I have admitted the high improbability, that if every fossil were disinterred, they would compose in each of the Divisions of Nature a perfect

[1] *Origin*, Ed. i. p. 329, vi. p. 471.

[2] The structure of the Pachyderm leg was a favourite with the author. It is discussed in the Essay of 1842, p. 48. In the present Essay the following sentence in the margin appears to refer to Pachyderms and Ruminants: "There can be no doubt, if we banish all fossils, existing groups stand more separate." The following occurs between the lines "The earliest forms would be such as others could radiate from."

series of the kind required; consequently I freely admit, that if those geologists are in the right who consider the lowest known formation as contemporaneous with the first appearances of life[1]; or the several formations as at all closely consecutive; or any one formation as containing a nearly perfect record of the organisms which existed during the whole period of its deposition in that quarter of the globe;—if such propositions are to be accepted, my theory must be abandoned.

If the Palæozoic system is really contemporaneous with the first appearance of life, my theory must be abandoned, both inasmuch as it limits *from shortness of time* the total number of forms which can have existed on this world, and because the organisms, as fish, mollusca[2] and star-fish found in its lower beds, cannot be considered as the parent forms of all the successive species in these classes. But no one has yet overturned the arguments of Hutton and Lyell, that the lowest formations known to us are only those which have escaped being metamorphosed ⟨illegible⟩; if we argued from some considerable districts, we might have supposed that even the Cretaceous system was that in which life first appeared. From the number of distant points, however, in which the Silurian system has been found to be the lowest, and not always metamorphosed, there are some objections to Hutton's and Lyell's view; but we must not forget that the now existing land forms only $\frac{1}{5}$ part of the superficies of the globe, and that this fraction is only imperfectly known. With respect to the fewness of the organisms found in the Silurian and other Palæozoic formations, there is less difficulty, inasmuch as

[1] *Origin*, Ed. i. p. 307, vi. p. 448.

[2] ⟨Pencil insertion by the author.⟩ The parent-forms of Mollusca would probably differ greatly from all recent,—it is not directly that any one division of Mollusca would descend from first time unaltered, whilst others had become metamorphosed from it.

(besides their gradual obliteration) we can expect formations of this vast antiquity to escape entire denudation, only when they have been accumulated over a wide area, and have been subsequently protected by vast superimposed deposits: now this could generally only hold good with deposits accumulating in a wide and deep ocean, and therefore unfavourable to the presence of many living things. A mere narrow and not very thick strip of matter, deposited along a coast where organisms most abound, would have no chance of escaping denudation and being preserved to the present time from such immensely distant ages[1].

If the several known formations are at all nearly consecutive in time, and preserve a fair record of the organisms which have existed, my theory must be abandoned. But when we consider the great changes in mineralogical nature and texture between successive formations, what vast and entire changes in the geography of the surrounding countries must generally have been effected, thus wholly to have changed the nature of the deposits on the same area. What time such changes must have required! Moreover how often has it not been found, that between two conformable and apparently immediately successive deposits a vast pile of water-worn matter is interpolated in an adjoining district. We have no means of conjecturing in many cases how long a period[2] has elapsed between successive formations, for the species are often wholly different: as remarked by Lyell, in some cases probably as long a period has elapsed between two formations as the whole Tertiary system, itself broken by wide gaps.

Consult the writings of any one who has particularly attended to any one stage in the Tertiary

[1] *Origin*, Ed. i. p. 291, vi. p. 426.

[2] ⟨Note in original.⟩ Reflect on coming in of the Chalk, extending from Iceland to the Crimea.

system (and indeed of every system) and see how deeply impressed he is with the time required for its accumulation[1]. Reflect on the years elapsed in many cases, since the latest beds containing only living species have been formed;—see what Jordan Smith says of the 20,000 years since the last bed, which is above the boulder formation in Scotland, has been upraised; or of the far longer period since the recent beds of Sweden have been upraised 400 feet, what an enormous period the boulder formation must have required, and yet how insignificant are the records (although there has been plenty of elevation to bring up submarine deposits) of the shells, which we know existed at that time. Think, then, over the entire length of the Tertiary epoch, and think over the probable length of the intervals, separating the Secondary deposits. Of these deposits, moreover, those consisting of sand and pebbles have seldom been favourable, either to the embedment or to the preservation of fossils[2].

Nor can it be admitted as probable that any one Secondary formation contains a fair record even of those organisms which are most easily preserved, namely hard marine bodies. In how many cases have we not certain evidence that between the deposition of apparently closely consecutive beds, the lower one existed for an unknown time as land, covered with trees. Some of the Secondary formations which contain most marine remains appear to have been formed in a wide and not deep sea, and therefore only those marine animals which live in such situations would be preserved[3]. In all cases, on indented rocky coasts, or any other coast, where sediment is not accumulating, although often highly

[1] *Origin*, Ed. i. p. 282, vi. p. 416.
[2] *Origin*, Ed. i. pp. 288, 300, vi. pp. 422, 438.
[3] ⟨Note in original.⟩ Neither highest or lowest fish (*i.e.* Myxina ⟨?⟩ or Lepidosiren) could be preserved in intelligible condition in fossils.

favourable to marine animals, none can be embedded: where pure sand and pebbles are accumulating few or none will be preserved. I may here instance the great western line of the S. American coast[1], tenanted by many peculiar animals, of which none probably will be preserved to a distant epoch. From these causes, and especially from such deposits as are formed along a line of coast, steep above and below water, being necessarily of little width, and therefore more likely to be subsequently denuded and worn away, we can see why it is improbable that our Secondary deposits contain a fair record of the Marine Fauna of any one period. The East Indian Archipelago offers an area, as large as most of our Secondary deposits, in which there are wide and shallow seas, teeming with marine animals, and in which sediment is accumulating; now supposing that all the hard marine animals, or rather those having hard parts to preserve, were preserved to a future age, excepting those which lived on rocky shores where no sediment or only sand and gravel were accumulating, and excepting those embedded along the steeper coasts, where only a narrow fringe of sediment was accumulating, supposing all this, how poor a notion would a person at a future age have of the Marine Fauna of the present day. Lyell[2] has compared the geological series to a work of which only the few latter but not consecutive chapters have been preserved; and out of which, it may be added, very many leaves have been torn, the remaining ones only illustrating a scanty portion of the Fauna of each period. On

[1] *Origin*, Ed. i. p. 290, vi. p. 425.

[2] See *Origin*, Ed. i. p. 310, vi. p. 452 for Lyell's metaphor. I am indebted to Prof. Judd for pointing out that Darwin's version of the metaphor is founded on the first edition of Lyell's *Principles*, vol. I. and vol. III.; see the Essay of 1842, p. 27.

this view, the records of anteceding ages confirm my theory; on any other they destroy it.

Finally, if we narrow the question into, why do we not find in some instances every intermediate form between any two species? the answer may well be that the average duration of each specific form (as we have good reason to believe) is immense in years, and that the transition could, according to my theory, be effected only by numberless small gradations; and therefore that we should require for this end a most perfect record, which the foregoing reasoning teaches us not to expect. It might be thought that in a vertical section of great thickness in the same formation some of the species ought to be found to vary in the upper and lower parts[1], but it may be doubted whether any formation has gone on accumulating without any break for a period as long as the duration of a species; and if it had done so, we should require a series of specimens from every part. How rare must be the chance of sediment accumulating for some 20 or 30 thousand years on the same spot[2], with the bottom subsiding, so that a proper depth might be preserved for any one species to continue living: what an amount of subsidence would be thus required, and this subsidence must not destroy the source whence the sediment continued to be derived. In the case of terrestrial animals, what chance is there when the present time is become a pleistocene formation (at an earlier period than this, sufficient elevation to expose marine beds could not be expected), what chance is there that future geologists will make out the innumerable transitional sub-varieties, through which the short-horned and long-

[1] See *More Letters*, vol. I. pp. 344–7, for Darwin's interest in the celebrated observations of Hilgendorf and Hyatt.

[2] This corresponds partly to *Origin*, Ed. i. p. 294, vi. p. 431.

horned cattle (so different in shape of body) have been derived from the same parent stock[1]? Yet this transition has been effected in *the same country*, and in a far *shorter time*, than would be probable in a wild state, both contingencies highly favourable for the future hypothetical geologists being enabled to trace the variation.

[1] *Origin*, Ed. i. p. 299, vi. p. 437.

CHAPTER V

GRADUAL APPEARANCE AND DISAPPEARANCE OF SPECIES[1]

In the Tertiary system, in the last uplifted beds, we find all the species recent and living in the immediate vicinity; in rather older beds we find only recent species, but some not living in the immediate vicinity[2]; we then find beds with two or three or a few more extinct or very rare species; then considerably more extinct species, but with gaps in the regular increase; and finally we have beds with only two or three or not one living species. Most geologists believe that the gaps in the percentage, that is the sudden increments, in the number of the extinct species in the stages of the Tertiary system are due to the imperfection of the geological record. Hence we are led to believe that the species in the Tertiary system have been gradually introduced; and from analogy to carry on the same view to the Secondary formations. In these latter, however, entire groups of species generally come in abruptly; but this would naturally result, if, as argued in the foregoing chapter, these Secondary deposits are separated by wide epochs. Moreover it is important to observe that, with our increase of knowledge, the gaps between the older formations become fewer and smaller; geologists of

[1] This chapter corresponds to ch. X of *Origin*, Ed. i., vi. ch. XI, "On the geological succession of organic beings."

[2] *Origin*, Ed. i. p. 312, vi. p. 453.

a few years standing remember how beautifully has the Devonian system[1] come in between the Carboniferous and Silurian formations. I need hardly observe that the slow and gradual appearance of new forms follows from our theory, for to form a new species, an old one must not only be plastic in its organization, becoming so probably from changes in the conditions of its existence, but a place in the natural economy of the district must [be made,] come to exist, for the selection of some new modification of its structure, better fitted to the surrounding conditions than are the other individuals of the same or other species[2].

In the Tertiary system the same facts, which make us admit as probable that new species have slowly appeared, lead to the admission that old ones have slowly disappeared, not several together, but one after another; and by analogy one is induced to extend this belief to the Secondary and Palæozoic epochs. In some cases, as the subsidence of a flat country, or the breaking or the joining of an isthmus, and the sudden inroad of many new and destructive species, extinction might be locally sudden. The view entertained by many geologists, that each fauna of each Secondary epoch has been suddenly destroyed over the whole world, so that no succession could be left for the production of new forms, is subversive of my theory, but I see no grounds whatever to admit such a view. On the

[1] In the margin the author has written "Lonsdale." This refers to W. Lonsdale's paper "Notes on the age of the Limestone of South Devonshire," *Geolog. Soc. Trans.*, Series 2, vol. v. 1840, p. 721. According to Mr H. B. Woodward (*History of the Geological Society of London*, 1907, p. 107) "Lonsdale's 'important and original suggestion of the existence of an intermediary type of Palæozoic fossils, since called Devonian,' led to a change which was then 'the greatest ever made at one time in the classification of our English formations'." Mr Woodward's quotations are from Murchison and Buckland.

[2] ⟨Note in original.⟩ Better begin with this. If species really, after catastrophes, created in showers over world, my theory false. ⟨In the above passage the author is obviously close to his theory of divergence.⟩

contrary, the law, which has been made out, with reference to distinct epochs, by independent observers, namely, that the wider the geographical range of a species the longer is its duration in time, seems entirely opposed to any universal extermination[1]. The fact of species of mammiferous animals and fish being renewed at a quicker rate than mollusca, though both aquatic; and of these the terrestrial genera being renewed quicker than the marine; and the marine mollusca being again renewed quicker than the Infusorial animalcula, all seem to show that the extinction and renewal of species does not depend on general catastrophes, but on the particular relations of the several classes to the conditions to which they are exposed[2].

Some authors seem to consider the fact of a few species having survived[3] amidst a number of extinct forms (as is the case with a tortoise and a crocodile out of the vast number of extinct sub-Himalayan fossils) as strongly opposed to the view of species being mutable. No doubt this would be the case, if it were presupposed with Lamarck that there was some inherent tendency to change and development in all species, for which supposition I see no evidence. As we see some species at present adapted to a wide range of conditions, so we may suppose that such species would survive unchanged and unexterminated for a long time; time generally being from geological causes a correlative of changing conditions. How at present one species becomes adapted to a wide range, and another species to a restricted range of conditions, is of difficult explanation.

[1] Opposite to this passage the author has written "d'Archiac, Forbes, Lyell."

[2] This passage, for which the author gives as authorities the names of Lyell, Forbes and Ehrenberg, corresponds in part to the discussion beginning on p. 313 of *Origin*, Ed. i., vi. p. 454.

[3] The author gives Falconer as his authority: see *Origin*, Ed. i. p. 313, vi. p. 454.

Extinction of species.

The extinction of the larger quadrupeds, of which we imagine we better know the conditions of existence, has been thought little less wonderful than the appearance of new species; and has, I think, chiefly led to the belief of universal catastrophes. When considering the wonderful disappearance within a late period, whilst recent shells were living, of the numerous great and small mammifers of S. America, one is strongly induced to join with the catastrophists. I believe, however, that very erroneous views are held on this subject. As far as is historically known, the disappearance of species from any one country has been slow—the species becoming rarer and rarer, locally extinct, and finally lost[1]. It may be objected that this has been effected by man's direct agency, or by his indirect agency in altering the state of the country; in this latter case, however, it would be difficult to draw any just distinction between his agency and natural agencies. But we now know in the later Tertiary deposits, that shells become rarer and rarer in the successive beds, and finally disappear: it has happened, also, that shells common in a fossil state, and thought to have been extinct, have been found to be still living species, but very *rare* ones[2]. If the rule is that organisms become extinct by becoming rarer and rarer, we ought not to view their extinction, even in the case of the larger quadrupeds, as anything wonderful and out of the common course of events. For no naturalist thinks it wonderful that one species of a genus should be rare and another abundant, notwithstanding he be

[1] This corresponds approximately to *Origin*, Ed. i. p. 317, vi. p. 458.

[2] The case of *Trigonia*, a great Secondary genus of shells surviving in a single species in the Australian seas, is given as an example in the *Origin*, Ed. i. p. 321, vi. p. 463.

quite incapable of explaining the causes of the comparative rareness[1]. Why is one species of willow-wren or hawk or woodpecker common in England, and another extremely rare: why at the Cape of Good Hope is one species of rhinoceros or antelope far more abundant than other species? Why again is the same species much more abundant in one district of a country than in another district? No doubt there are in each case good causes: but they are unknown and unperceived by us. May we not then safely infer that as certain causes are acting *unperceived* around us, and are making one species to be common and another exceedingly rare, that they might equally well cause the final extinction of some species without being perceived by us? We should always bear in mind that there is a recurrent struggle for life in every organism, and that in every country a destroying agency is always counteracting the geometrical tendency to increase in every species; and yet without our being able to tell with certainty at what period of life, or at what period of the year, the destruction falls the heaviest. Ought we then to expect to trace the steps by which this destroying power, always at work and scarcely perceived by us, becomes increased, and yet if it continues to increase ever so slowly (without the fertility of the species in question be likewise increased) the average number of the individuals of that species must decrease, and become finally lost. I may give a single instance of a check causing local extermination which might long have escaped discovery[2]; the horse, though swarming in a wild state in La Plata, and likewise under apparently the most unfavourable conditions in the scorched and alternately flooded plains of Caraccas, will not in a wild

[1] This point, on which the author laid much stress, is discussed in the *Origin*, Ed. i. p. 319, vi. p. 461.

[2] *Origin*, Ed. i. p. 72, vi. p. 89.

state extend beyond a certain degree of latitude into the intermediate country of Paraguay; this is owing to a certain fly depositing its eggs on the navels of the foals: as, however, man with a *little* care can rear horses in a tame state *abundantly* in Paraguay, the problem of its extinction is probably complicated by the greater exposure of the wild horse to occasional famine from the droughts, to the attacks of the jaguar and other such evils. In the Falkland Islands the check to the *increase* of the wild horse is said to be loss of the sucking foals[1], from the stallions compelling the mares to travel across bogs and rocks in search of food: if the pasture on these islands decreased a little, the horse, perhaps, would cease to exist in a wild state, not from the absolute want of food, but from the impatience of the stallions urging the mares to travel whilst the foals were too young.

From our more intimate acquaintance with domestic animals, we cannot conceive their extinction without some glaring agency; we forget that they would undoubtedly in a state of nature (where other animals are ready to fill up their place) be acted on in some part of their lives by a destroying agency, keeping their numbers on an average constant. If the common ox was known only as a wild S. African species, we should feel no surprise at hearing that it was a very rare species; and this rarity would be a stage towards its extinction. Even in man, so infinitely better known than any other inhabitant of this world, how impossible it has been found, without statistical calculations, to judge of the proportions of births and deaths, of the duration of life, and of the increase and decrease of population; and still less of the causes of such changes: and yet, as has so often been repeated, decrease in

[1] This case does not occur in the *Origin*, Ed.

numbers or rarity seems to be the high-road to extinction. To marvel at the extermination of a species appears to me to be the same thing as to know that illness is the road to death,—to look at illness as an ordinary event, nevertheless to conclude, when the sick man dies, that his death has been caused by some unknown and violent agency[1].

In a future part of this work we shall show that, as a general rule, groups of allied species[2] gradually appear and disappear, one after the other, on the face of the earth, like the individuals of the same species: and we shall then endeavour to show the probable cause of this remarkable fact.

[1] An almost identical sentence occurs in the *Origin*, Ed. i. p. 320, vi. p. 462.

[2] *Origin*, Ed. i. p. 316, vi. p. 457.

CHAPTER VI

ON THE GEOGRAPHICAL DISTRIBUTION OF ORGANIC BEINGS IN PAST AND PRESENT TIMES

FOR convenience sake I shall divide this chapter into three sections[1]. In the first place I shall endeavour to state the laws of the distribution of existing beings, as far as our present object is concerned; in the second, that of extinct; and in the third section I shall consider how far these laws accord with the theory of allied species having a common descent.

SECTION FIRST.

Distribution of the inhabitants in the different continents.

In the following discussion I shall chiefly refer to terrestrial mammifers, inasmuch as they are better known; their differences in different countries, strongly marked; and especially as the necessary

[1] Chapters XI and XII in the *Origin*, Ed. i., vi. chs. XII and XIII ("On geographical distribution") show signs of having been originally one, in the fact that one summary serves for both. The geological element is not separately treated there, nor is there a separate section on "how far these laws accord with the theory, &c."

In the MS. the author has here written in the margin "If same species appear at two spot at once, fatal to my theory." See *Origin*, Ed. i. p. 352, vi. p. 499

means of their transport are more evident, and confusion, from the accidental conveyance by man of a species from one district to another district, is less likely to arise. It is known that all mammifers (as well as all other organisms) are united in one great system; but that the different species, genera, or families of the same order inhabit different quarters of the globe. If we divide the land[1] into two divisions, according to the amount of difference, and disregarding the numbers of the terrestrial mammifers inhabiting them, we shall have first Australia including New Guinea; and secondly the rest of the world: if we make a three-fold division, we shall have Australia, S. America, and the rest of the world; I must observe that North America is in some respects neutral land, from possessing some S. American forms, but I believe it is more closely allied (as it certainly is in its birds, plants and shells) with Europe. If our division had been four-fold, we should have had Australia, S. America, Madagascar (though inhabited by few mammifers) and the remaining land: if five-fold, Africa, especially the southern eastern parts, would have to be separated from the remainder of the world. These differences in the mammiferous inhabitants of the several main divisions of the globe cannot, it is well known, be explained by corresponding differences in their conditions[2]; how similar are parts of tropical America and Africa; and accordingly we find some *analogous* resemblances,—thus both have monkeys, both large feline animals, both large Lepidoptera, and large dung-feeding beetles; both have palms and epiphytes; and yet the essential difference between their productions is as great as between those of the arid plains of the Cape of Good Hope

[1] This division of the land into regions does not occur in the *Origin*, Ed. i.

[2] *Origin*, Ed. i. p. 346, vi. p. 493.

and the grass-covered savannahs of La Plata[1]. Consider the distribution of the Marsupialia, which are eminently characteristic of Australia, and in a lesser degree of S. America; when we reflect that animals of this division, feeding both on animal and vegetable matter, frequent the dry open or wooded plains and mountains of Australia, the humid impenetrable forests of New Guinea and Brazil; the dry rocky mountains of Chile, and the grassy plains of Banda Oriental, we must look to some other cause, than the nature of the country, for their absence in Africa and other quarters of the world.

Furthermore it may be observed that *all* the organisms inhabiting any country are not perfectly adapted to it[2]; I mean by not being perfectly adapted, only that some few other organisms can generally be found better adapted to the country than some of the aborigines. We must admit this when we consider the enormous number of horses and cattle which have run wild during the three last centuries in the uninhabited parts of St Domingo, Cuba, and S. America; for these animals must have supplanted some aboriginal ones. I might also adduce the same fact in Australia, but perhaps it will be objected that 30 or 40 years has not been a sufficient period to test this power of struggling ⟨with⟩ and overcoming the aborigines. We know the European mouse is driving before it that of New Zealand, like the Norway rat has driven before it the old English species in England. Scarcely an island can be named, where casually introduced plants have not supplanted some of the native species: in La Plata the Cardoon covers square leagues of country on

[1] Opposite this passage is written "*not botanically*," in Sir J. D. Hooker's hand. The word *palms* is underlined three times and followed by three exclamation marks. An explanatory note is added in the margin "singular paucity of palms and epiphytes in Trop. Africa compared with Trop. America and Ind. Or." ⟨=East Indies⟩.

[2] This partly corresponds to *Origin*, Ed. i. p. 337, vi. p. 483.

which some S. American plants must once have grown: the commonest weed over the whole of India is an introduced Mexican poppy. The geologist who knows that slow changes are in progress, replacing land and water, will easily perceive that even if all the organisms of any country had originally been the best adapted to it, this could hardly continue so during succeeding ages without either extermination, or changes, first in the relative proportional numbers of the inhabitants of the country, and finally in their constitutions and structure.

Inspection of a map of the world at once shows that the five divisions, separated according to the greatest amount of difference in the mammifers inhabiting them, are likewise those most widely separated from each other by barriers[1] which mammifers cannot pass: thus Australia is separated from New Guinea and some small adjoining islets only by a narrow and shallow strait; whereas New Guinea and its adjoining islets are cut off from the other East Indian islands by deep water. These latter islands, I may remark, which fall into the great Asiatic group, are separated from each other and the continent only by shallow water; and where this is the case we may suppose, from geological oscillations of level, that generally there has been recent union. South America, including the southern part of Mexico, is cut off from North America by the West Indies, and the great tableland of Mexico, except by a mere fringe of tropical forests along the coast: it is owing, perhaps, to this fringe that N. America possesses some S. American forms. Madagascar is entirely isolated. Africa is also to a great extent isolated, although it approaches, by many promontories and by lines of shallower sea, to Europe and Asia: southern Africa, which is

[1] On the general importance of barriers, see *Origin*, Ed. i. p. 347, vi. p. 494.

the most distinct in its mammiferous inhabitants, is separated from the northern portion by the Great Sahara Desert and the table-land of Abyssinia. That the distribution of organisms is related to barriers, stopping their progress, we clearly see by comparing the distribution of marine and terrestrial productions. The marine animals being different on the two sides of land tenanted by the same terrestrial animals, thus the shells are wholly different on the opposite sides of the temperate parts of South America[1], as they are (?) in the Red Sea and the Mediterranean. We can at once perceive that the destruction of a barrier would permit two geographical groups of organisms to fuse and blend into one. But the original cause of groups being different on opposite sides of a barrier can only be understood on the hypothesis of each organism having been created or produced on one spot or area, and afterwards migrating as widely as its means of transport and subsistence permitted it.

Relation of range in genera and species.

It is generally[2] found, that where a genus or group ranges over nearly the entire world, many of the species composing the group have wide ranges: on the other hand, where a group is restricted to any one country, the species composing it generally have restricted ranges in that country[3]. Thus among mammifers the feline and canine genera are widely distributed, and many of the individual species have enormous ranges [the genus Mus I believe, however, is a strong exception to the rule].

[1] *Origin*, Ed. i. p. 348, vi. p. 495.

[2] ⟨Note in original.⟩ The same laws seem to govern distribution of species and genera, and individuals in time and space. ⟨See *Origin*, Ed. i. p. 350, vi. p. 497, also a passage in the last chapter, p. 146.⟩

[3] *Origin*, Ed. i. p. 404, vi. p. 559.

Mr Gould informs me that the rule holds with birds, as in the owl genus, which is mundane, and many of the species range widely. The rule holds also with land and fresh-water mollusca, with butterflies and very generally with plants. As instances of the converse rule, I may give that division of the monkeys which is confined to S. America, and amongst plants, the Cacti, confined to the same continent, the species of both of which have generally narrow ranges. On the ordinary theory of the separate creation of each species, the cause of these relations is not obvious; we can see no reason, because many allied species have been created in the several main divisions of the world, that several of these species should have wide ranges; and on the other hand, that species of the same group should have narrow ranges if all have been created in one main division of the world. As the result of such and probably many other unknown relations, it is found that, even in the same great classes of beings, the different divisions of the world are characterised by either merely different species, or genera, or even families: thus in cats, mice, foxes, S. America differs from Asia and Africa only in species; in her pigs, camels and monkeys the difference is generic or greater. Again, whilst southern Africa and Australia differ more widely in their mammalia than do Africa and S. America, they are more closely (though indeed very distantly) allied in their plants.

Distribution of the inhabitants in the same continent.

If we now look at the distribution of the organisms in any one of the above main divisions of the world, we shall find it split up into many regions, with all or nearly all their species distinct, but yet partaking of one common character. This similarity of type in the subdivisions of a great region is equally well-known with the dissimilarity of the inhabitants of the several great regions; but it has been less often insisted on, though more worthy of remark. Thus for instance, if in Africa or S. America, we go from south to north[1], or from lowland to upland, or from a humid to a dryer part, we find wholly different species of those genera or groups which characterise the continent over which we are passing. In these subdivisions we may clearly observe, as in the main divisions of the world, that sub-barriers divide different groups of species, although the opposite sides of such sub-barriers may possess nearly the same climate, and may be in other respects nearly similar: thus it is on the opposite sides of the Cordillera of Chile, and in a lesser degree on the opposite sides of the Rocky mountains. Deserts, arms of the sea, and even rivers form the barriers; mere preoccupied space seems sufficient in several cases: thus Eastern and Western Australia, in the same latitude, with very similar climate and soils, have scarcely a plant, and few animals or birds, in common, although all belong to the peculiar genera characterising Australia. It is in short impossible to explain the differences in the inhabitants, either of the main divisions of the world, or of these sub-divisions, by the differences in their physical conditions, and by the adaptation of their inhabitants. Some other cause must intervene.

We can see that the destruction of sub-barriers would cause (as before remarked in the case of the main divisions) two sub-divisions to blend into one; and we can only suppose that the original difference in the species, on the opposite sides of sub-barriers, is due to the creation or production of

[1] *Origin*, Ed. i. p. 349, vi. p. 496.

species in distinct areas, from which they have wandered till arrested by such sub-barriers. Although thus far is pretty clear, it may be asked, why, when species in the same main division of the world were produced on opposite sides of a sub-barrier, both when exposed to similar conditions and when exposed to widely different influences (as on alpine and lowland tracts, as on arid and humid soils, as in cold and hot climates), have they invariably been formed on a similar type, and that type confined to this one division of the world? Why when an ostrich[1] was produced in the southern parts of America, was it formed on the American type, instead of on the African or on Australian types? Why when hare-like and rabbit-like animals were formed to live on the Savannahs of La Plata, were they produced on the peculiar Rodent type of S. America, instead of on the true[2] hare-type of North America, Asia and Africa? Why when borrowing Rodents, and camel-like animals were formed to tenant the Cordillera, were they formed on the same type[3] with their representatives on the plains? Why were the mice, and many birds of different species on the opposite sides of the Cordillera, but exposed to a very similar climate and soil, created on the same peculiar S. American type? Why were the plants in Eastern and Western Australia, though wholly different as species, formed on the same peculiar Australian types? The generality of the rule, in so many places and under such different circumstances, makes it highly remarkable and seems to demand some explanation.

[1] The case of the ostrich (*Rhea*) occurs in the *Origin*, Ed. i. p. 349, vi. p. 496.

[2] ⟨Note in original.⟩ There is a hare in S. America,—so bad example.

[3] See *Origin*, Ed. i. p. 349, vi. p. 497.

Insular Faunas.

If we now look to the character of the inhabitants of small islands[1], we shall find that those situated close to other land have a similar fauna with that land[2], whilst those at a considerable distance from other land often possess an almost entirely peculiar fauna. The Galapagos Archipelago[3] is a remarkable instance of this latter fact; here almost every bird, its one mammifer, its reptiles, land and sea shells, and even fish, are almost all peculiar and distinct species, not found in any other quarter of the world: so are the majority of its plants. But although situated at the distance of between 500 and 600 miles from the S. American coast, it is impossible to even glance at a large part of its fauna, especially at the birds, without at once seeing that they belong to the American type[4]. Hence, in fact, groups of islands thus circumstanced form merely small but well-defined sub-divisions of the larger geographical divisions. But the fact is in such cases far more striking: for taking the Galapagos Archipelago as an instance; in the first place we must feel convinced, seeing that every island is wholly volcanic and bristles with craters, that in a geological sense the whole is of recent origin comparatively with a continent; and as the species are nearly all peculiar, we must conclude that they have in the same sense recently been produced on this very spot; and

[1] For the general problem of Oceanic Islands, see *Origin*, Ed. i. p. 388, vi. p. 541.

[2] This is an illustration of the general theory of barriers (*Origin*, Ed. i. p. 347, vi. p. 494). At i. p. 391, vi. p. 544 the question is discussed from the point of view of means of transport. Between the lines, above the words "with that land," the author wrote "Cause, formerly joined, no one doubts after Lyell."

[3] *Origin*, Ed. i. p. 390, vi. p. 543.

[4] See *Origin*, Ed. i. p. 397, vi. p. 552.

although in the nature of the soil, and in a lesser degree in the climate, there is a wide difference with the nearer part of the S. American coast, we see that the inhabitants have been formed on the same closely allied type. On the other hand, these islands, as far as their physical conditions are concerned, resemble closely the Cape de Verde volcanic group, and yet how wholly unlike are the productions of these two archipelagoes. The Cape de Verde[1] group, to which may be added the Canary Islands, are allied in their inhabitants (of which many are peculiar species) to the coast of Africa and southern Europe, in precisely the same manner as the Galapagos Archipelago is allied to America. We here clearly see that mere geographical proximity affects, more than any relation of adaptation, the character of species. How many islands in the Pacific exist far more like in their physical conditions to Juan Fernandez than this island is to the coast of Chile, distant 300 miles; why then, except from mere proximity, should this island alone be tenanted by two very peculiar species of humming-birds—that form of birds which is so exclusively American? Innumerable other similar cases might be adduced.

The Galapagos Archipelago offers another, even more remarkable, example of the class of facts we are here considering. Most of its genera are, as we have said, American, many of them are mundane, or found everywhere, and some are quite or nearly confined to this archipelago. The islands are of absolutely similar composition, and exposed to the same climate; most of them are in sight of each other; and yet several of the islands are inhabited, each by peculiar species (or in some cases perhaps only varieties) of some of the genera characterising the archipelago. So that the small group of the Gala-

[1] The Cape de Verde and Galapagos Archipelagoes are compared in the *Origin*, Ed. i. p. 398, vi. p. 553. See also *Journal of Researches*, 1860, p. 393.

pagos Islands typifies, and follows exactly the same laws in the distribution of its inhabitants, as a great continent. How wonderful it is that two or three closely similar but distinct species of a mocking-thrush[1] should have been produced on three neighbouring and absolutely similar islands; and that these three species of mocking-thrush should be closely related to the other species inhabiting wholly different climates and different districts of America, and only in America. No similar case so striking as this of the Galapagos Archipelago has hitherto been observed; and this difference of the productions in the different islands may perhaps be partly explained by the depth of the sea between them (showing that they could not have been united within recent geological periods), and by the currents of the sea sweeping *straight* between them,—and by storms of wind being rare, through which means seeds and birds could be blown, or drifted, from one island to another. There are however some similar facts: it is said that the different, though neighbouring islands of the East Indian Archipelago are inhabited by some different species of the same genera; and at the Sandwich group some of the islands have each their peculiar species of the same genera of plants.

Islands standing quite isolated within the intratropical oceans have generally very peculiar floras, related, though feebly (as in the case of St Helena[2] where almost every species is distinct), with the nearest continent: Tristan d'Acunha is feebly related, I believe, in its plants, both to Africa and S. America, not by having species in common, but

[1] In the *Origin*, Ed. i. p. 390, a strong point is made of birds which immigrated "with facility and in a body" not having been modified. Thus the author accounts for the small percentage of peculiar "marine birds."

[2] "The affinities of the St Helena flora are strongly South African." Hooker's *Lecture on Insular Floras* in the *Gardeners' Chronicle*, Jan. 1867.

by the genera to which they belong[1]. The floras of the numerous scattered islands of the Pacific are related to each other and to all the surrounding continents; but it has been said, that they have more of an Indo-Asiatic than American character[2]. This is somewhat remarkable, as America is nearer to all the Eastern islands, and lies in the direction of the trade-wind and prevailing currents; on the other hand, all the heaviest gales come from the Asiatic side. But even with the aid of these gales, it is not obvious on the ordinary theory of creation how the possibility of migration (without we suppose, with extreme improbability, that each species with an Indo-Asiatic character has actually travelled from the Asiatic shores, where such species do not now exist) explains this Asiatic character in the plants of the Pacific. This is no more obvious than that (as before remarked) there should exist a relation between the creation of closely allied species in several regions of the world, and the fact of many such species having wide ranges; and on the other hand, of allied species confined to one region of the world having in that region narrow ranges.

Alpine Floras.

We will now turn to the floras of mountain-summits which are well known to differ from the floras of the neighbouring lowlands. In certain characters, such as dwarfness of stature, hairiness, &c., the species from the most distant mountains frequently resemble each other,—a kind of analogy like that for instance of the succulency of most desert plants. Besides this analogy, Alpine plants

[1] It is impossible to make out the precise form which the author intended to give to this sentence, but the meaning is clear.

[2] This is no doubt true, the flora of the Sandwich group however has marked American affinities.

present some eminently curious facts in their distribution. In some cases the summits of mountains, although immensely distant from each other, are clothed by the same identical species[1] which are likewise the same with those growing on the likewise very distant Arctic shores. In other cases, although few or none of the species may be actually identical, they are closely related; whilst the plants of the lowland districts surrounding the two mountains in question will be wholly dissimilar. As mountain-summits, as far as their plants are concerned, are islands rising out of an ocean of land in which the Alpine species cannot live, nor across which is there any known means of transport, this fact appears directly opposed to the conclusion which we have come to from considering the general distribution of organisms both on continents and on islands—namely, that the degree of relationship between the inhabitants of two points depends on the completeness and nature of the barriers between those points[2]. I believe, however, this anomalous case admits, as we shall presently see, of some explanation. We might have expected that the flora of a mountain summit would have presented the same relation to the flora of the surrounding lowland country, which any isolated part of a continent does to the whole, or an island does to the mainland, from which it is separated by a rather wide space of sea. This in fact is the case with the plants clothing the summits of *some* mountains, which mountains it may be observed are particularly isolated; for instance, all the species are peculiar, but they belong to the forms characteristic of the surrounding continent, on the mountains of Caraccas, of Van Dieman's

[1] See *Origin*, Ed. i. p. 365, vi. p. 515. The present discussion was written before the publication of Forbes' celebrated paper on the same subject; see *Life and Letters*, vol. I. p. 88.

[2] The apparent breakdown of the doctrine of barriers is slightly touched on in the *Origin*, Ed. i. p. 365, vi. p. 515.

Land and of the Cape of Good Hope[1]. On some other mountains, for instance ⟨in⟩ Tierra del Fuego and in Brazil, some of the plants though distinct species are S. American forms; whilst others are allied to or are identical with the Alpine species of Europe. In islands of which the lowland flora is distinct ⟨from⟩ but allied to that of the nearest continent, the Alpine plants are sometimes (or perhaps mostly) eminently peculiar and distinct[2]; this is the case on Teneriffe, and in a lesser degree even on some of the Mediterranean islands.

If all Alpine floras had been characterised like that of the mountain of Caraccas, or of Van Dieman's Land, &c., whatever explanation is possible of the general laws of geographical distribution would have applied to them. But the apparently anomalous case just given, namely of the mountains of Europe, of some mountains in the United States (Dr Boott) and of the summits of the Himalaya (Royle), having many identical species in common conjointly with the Arctic regions, and many species, though not identical, closely allied, require a separate explanation. The fact likewise of several of the species on the mountains of Tierra del Fuego (and in a lesser degree on the mountains of Brazil) not belonging to American forms, but to those of Europe, though so immensely remote, requires also a separate explanation.

[1] In the *Origin*, Ed. i. p. 375, vi. p. 526, the author points out that on the mountains at the Cape of Good Hope "some few representative European forms are found, which have not been discovered in the intertropical parts of Africa."

[2] See Hooker's *Lecture on Insular Floras* in the *Gardeners' Chronicle*, Jan. 1867.

Cause of the similarity in the floras of some distant mountains.

Now we may with confidence affirm, from the number of the then floating icebergs and low descent of the glaciers, that within a period so near that species of shells have remained the same, the whole of Central Europe and of North America (and perhaps of Eastern Asia) possessed a very cold climate; and therefore it is probable that the floras of these districts were the same as the present Arctic one,—as is known to have been to some degree the case with then existing sea-shells, and those now living on the Arctic shores. At this period the mountains must have been covered with ice of which we have evidence in the surfaces polished and scored by glaciers. What then would be the natural and almost inevitable effects of the gradual change into the present more temperate climate[1]? The ice and snow would disappear from the mountains, and as new plants from the more temperate regions of the south migrated northward, replacing the Arctic plants, these latter would crawl[2] up the now uncovered mountains, and likewise be driven northward to the present Arctic shores. If the Arctic flora of that period was a nearly uniform one, as the present one is, then we should have the same plants on these mountain-summits and on the present Arctic shores. On this view the Arctic flora of that period must have been a widely extended one, more so than even the present one; but considering how similar the physical conditions must always be of land bordering on perpetual frost, this does not appear a great difficulty; and may we not venture to suppose that

[1] In the margin the author has written "(Forbes)." This may have been inserted at a date later than 1844, or it may refer to a work by Forbes earlier than his Alpine paper.

[2] See *Origin*, Ed. i. p. 367, vi. p. 517.

the almost infinitely numerous icebergs, charged with great masses of rocks, soil and *brushwood*[1] and often driven high up on distant beaches, might have been the means of widely distributing the seeds of the same species?

I will only hazard one other observation, namely that during the change from an extremely cold climate to a more temperate one the conditions, both on lowland and mountain, would be singularly favourable for the diffusion of any existing plants, which could live on land, just freed from the rigour of eternal winter; for it would possess no inhabitants; and we cannot doubt that *preoccupation*[2] is the chief bar to the diffusion of plants. For amongst many other facts, how otherwise can we explain the circumstance that the plants on the opposite, though similarly constituted sides of a wide river in Eastern Europe (as I was informed by Humboldt) should be widely different; across which river birds, swimming quadrupeds and the wind must often transport seeds; we can only suppose that plants already occupying the soil and freely seeding check the germination of occasionally transported seeds.

At about the same period when icebergs were transporting boulders in N. America as far as 36° south, where the cotton tree now grows in South America, in latitude 42° (where the land is now clothed with forests having an almost tropical aspect with the trees bearing epiphytes and intertwined with canes), the same ice action was going on; is it not then in some degree probable that at this period the whole tropical parts of the two Americas

[1] ⟨Note in original.⟩ Perhaps vitality checked by cold and so prevented germinating. ⟨On the carriage of seeds by icebergs, see *Origin*, Ed. i. p. 363, vi. p. 513.⟩

[2] A note by the author gives "many authors" apparently as authority for this statement.

possessed[1] (as Falconer asserts that India did) a more temperate climate? In this case the Alpine plants of the long chain of the Cordillera would have descended much lower and there would have been a broad high-road[2] connecting those parts of North and South America which were then frigid. As the present climate supervened, the plants occupying the districts which now are become in both hemispheres temperate and even semi-tropical must have been driven to the Arctic and Antarctic[3] regions; and only a few of the loftiest points of the Cordillera can have retained their former connecting flora. The transverse chain of Chiquitos might perhaps in a similar manner during the ice-action period have served as a connecting road (though a broken one) for Alpine plants to become dispersed from the Cordillera to the highlands of Brazil. It may be observed that some (though not strong) reasons can be assigned for believing that at about this same period the two Americas were not so thoroughly divided as they now are by the West Indies and tableland of Mexico. I will only further remark that the present most singularly close similarity in the vegetation of the lowlands of Kerguelen's Land[4] and of Tierra del Fuego (Hooker), though so far apart, may perhaps be explained by the dissemination of seeds during this same cold period, by means of icebergs, as before alluded to[5].

Finally, I think we may safely grant from the foregoing facts and reasoning that the anomalous

[1] Opposite to this passage, in the margin, the author has written:—"too hypothetical."

[2] The Cordillera is described as supplying a great line of invasion in the *Origin*, Ed. i. p. 378.

[3] This is an approximation to the author's views on trans-tropical migration (*Origin*, Ed. i. pp. 376–8). See Thiselton-Dyer's interesting discussion in *Darwin and Modern Science*, p. 304.

[4] See Hooker's *Lecture on Insular Floras* in the *Gardeners' Chronicle*, Jan. 1867.

[5] ⟨Note by the author.⟩ Similarity of flora of coral islands easily explained.

similarity in the vegetation of certain very distant mountain-summits is not in truth opposed to the conclusion of the intimate relation subsisting between proximity in space (in accordance with the means of transport in each class) and the degree of affinity of the inhabitants of any two countries. In the case of several quite isolated mountains, we have seen that the general law holds good.

Whether the same species has been created more than once.

As the fact of the same species of plants having been found on mountain-summits immensely remote has been one chief cause of the belief of some species having been contemporaneously produced or created at two different points[1], I will here briefly discuss this subject. On the ordinary theory of creation, we can see no reason why on two similar mountain-summits two similar species may not have been created; but the opposite view, independently of its simplicity, has been generally received from the analogy of the general distribution of all organisms, in which (as shown in this chapter) we almost always find that great and continuous barriers separate distinct series; and we are naturally led to suppose that the two series have been separately created. When taking a more limited view we see a river, with a quite similar country on both sides, with one side well stocked with a certain animal and on the other side not one (as is the case with the Bizcacha[2] on the opposite sides of the Plata), we are at once led to conclude that the Bizcacha

[1] On centres of creation see *Origin*, Ed. i. p. 352, vi. p. 499.

[2] In the *Journal of Researches*, Ed. 1860, p. 124, the distribution of the Bizcacha is described as limited by the river Uruguay. The case is not I think given in the *Origin*.

was produced on some one point or area on the western side of the river. Considering our ignorance of the many strange chances of diffusion by birds (which occasionally wander to immense distances) and quadrupeds swallowing seeds and ova (as in the case of the flying water-beetle which disgorged the eggs of a fish), and of whirlwinds carrying seeds and animals into strong upper currents (as in the case of volcanic ashes and showers of hay, grain and fish[1]), and of the possibility of species having survived for short periods at intermediate spots and afterwards becoming extinct there[2]; and considering our knowledge of the great changes which *have* taken place from subsidence and elevation in the surface of the earth, and of our ignorance of the greater changes which *may have* taken place, we ought to be very slow in admitting the probability of double creations. In the case of plants on mountain-summits, I think I have shown how almost necessarily they would, under the past conditions of the northern hemisphere, be as similar as are the plants on the present Arctic shores; and this ought to teach us a lesson of caution.

But the strongest argument against double creations may be drawn from considering the case of mammifers[3] in which, from their nature and from the size of their offspring, the means of distribution are more in view. There are no cases where the same species is found in *very remote* localities,

[1] In the *Origin*, Ed. i. a special section (p. 356, vi. p. 504) is devoted to *Means of Dispersal.* The much greater prominence given to this subject in the *Origin* is partly accounted for by the author's experiments being of later date, *i.e.* 1855 (*Life and Letters*, vol. II. p. 53). The carriage of fish by whirlwinds is given in the *Origin*, Ed. i. p. 384, vi. p. 536.

[2] The case of islands serving as halting places is given in the *Origin*, Ed. i. p. 357, vi. p. 505. But here the evidence of this having occurred is supposed to be lost by the subsidence of the islands, not merely by the extinction of the species.

[3] "We find no inexplicable cases of the same mammal inhabiting distant points of the world." *Origin*, Ed. i. p. 352, vi. p. 500. See also *Origin*, Ed. i. p. 393, vi. p. 547.

except where there is a continuous belt of land: the Arctic region perhaps offers the strongest exception, and here we know that animals are transported on icebergs[1]. The cases of lesser difficulty may all receive a more or less simple explanation; I will give only one instance; the nutria[2], I believe, on the eastern coast of S. America live exclusively in fresh-water rivers, and I was much surprised how they could have got into rivulets, widely apart, on the coast of Patagonia; but on the opposite coast I found these quadrupeds living exclusively in the sea, and hence their migration along the Patagonian coast is not surprising. There is no case of the same mammifer being found on an island far from the coast, and on the mainland, as happens with plants[3]. On the idea of double creations it would be strange if the same species of several plants should have been created in Australia and Europe; and no one instance of the same species of mammifer having been created, or aboriginally existing, in two as nearly remote and equally isolated points. It is more philosophical, in such cases, as that of some plants being found in Australia and Europe, to admit that we are ignorant of the means of transport. I will allude only to one other case, namely, that of the Mydas[4], an Alpine animal, found only on the distant peaks of the mountains of Java: who will pretend to deny that during the ice period

[1] ⟨Note by the author.⟩ Many authors. ⟨See *Origin*, Ed. i. p. 394, vi. p. 547.⟩

[2] *Nutria* is the Spanish for otter, and is now a synonym for *Lutra*. The otter on the Atlantic coast is distinguished by minute differences from the Pacific species. Both forms are said to take to the sea. In fact the case presents no especial difficulties.

[3] In *Origin*, Ed. i. p. 394, vi. p. 548, bats are mentioned as an explicable exception to this statement.

[4] This reference is doubtless to *Mydaus*, a badger-like animal from the mountains of Java and Sumatra (Wallace, *Geographical Distribution*, ii. p. 199). The instance does not occur in the *Origin* but the author remarks (*Origin*, Ed. i. p. 376, vi. p. 527) that cases, strictly analogous to the distribution of plants, occur among terrestrial mammals.

of the northern and southern hemispheres, and when India is believed to have been colder, the climate might not have permitted this animal to haunt a lower country, and thus to have passed along the ridges from summit to summit? Mr Lyell has further observed that, *as in space, so in time,* there is no reason to believe that after the extinction of a species, the self-same form has ever reappeared[1]. I think, then, we may, notwithstanding the many cases of difficulty, conclude with some confidence that every species has been created or produced on a single point or area.

On the number of species, and of the classes to which they belong in different regions.

The last fact in geographical distribution, which, as far as I can see, in any way concerns the origin of species, relates to the absolute number and nature of the organic beings inhabiting different tracts of land. Although every species is admirably adapted (but not necessarily better adapted than every other species, as we have seen in the great increase of introduced species) to the country and station it frequents; yet it has been shown that the entire difference between the species in distant countries cannot possibly be explained by the difference of the physical conditions of these countries. In the same manner, I believe, neither the number of the species, nor the nature of the great classes to which they belong, can possibly in all cases be explained by the conditions of their country. New Zealand[2], a linear island stretching over about 700 miles of latitude, with forests, marshes, plains and mountains reaching to the limits of eternal snow, has far more

[1] See *Origin,* Ed. i. p. 313, vi. p. 454.

[2] The comparison between New Zealand and the Cape is given in the *Origin,* Ed. i. p. 389, vi. p. 542.

diversified habitats than an equal area at the Cape of Good Hope; and yet, I believe, at the Cape of Good Hope there are, of phanerogamic plants, from five to ten times the number of species as in all New Zealand. Why on the theory of absolute creations should this large and diversified island only have from 400 to 500 (?Dieffenbach) phanerogamic plants? and why should the Cape of Good Hope, characterised by the uniformity of its scenery, swarm with more species of plants than probably any other quarter of the world? Why on the ordinary theory should the Galapagos Islands abound with terrestrial reptiles? and why should many equal-sized islands in the Pacific be without a single one[1] or with only one or two species? Why should the great island of New Zealand be without one mammiferous quadruped except the mouse, and that was probably introduced with the aborigines? Why should not one island (it can be shown, I think, that the mammifers of Mauritius and St Iago have all been introduced) in the open ocean possess a mammiferous quadruped? Let it not be said that quadrupeds cannot live in islands, for we know that cattle, horses and pigs during a long period have run wild in the West Indian and Falkland Islands; pigs at St Helena; goats at Tahiti; asses in the Canary Islands; dogs in Cuba; cats at Ascension; rabbits at Madeira and the Falklands; monkeys at St Iago and the Mauritius; even elephants during a long time in one of the very small Sooloo Islands; and European mice on very many of the smallest islands far from the habitations of man[2]. Nor let it be assumed that quadrupeds are more slowly created and hence that the oceanic islands, which generally

[1] In a corresponding discussion in the *Origin*, Ed. i. p. 393, vi. p. 546, stress is laid on the distribution of Batrachians not of reptiles.

[2] The whole argument is given—more briefly than here—in the *Origin*, Ed. i. p. 394, vi. p. 547.

are of volcanic formation, are of too recent origin to possess them; for we know (Lyell) that new forms of quadrupeds succeed each other quicker than Mollusca or Reptilia. Nor let it be assumed (though such an assumption would be no explanation) that quadrupeds cannot be created on small islands; for islands not lying in mid-ocean do possess their peculiar quadrupeds; thus many of the smaller islands of the East Indian Archipelago possess quadrupeds; as does Fernando Po on the West Coast of Africa; as the Falkland Islands possess a peculiar wolf-like fox[1]; so do the Galapagos Islands a peculiar mouse of the S. American type. These two last are the most remarkable cases with which I am acquainted; inasmuch as the islands lie further from other land. It is possible that the Galapagos mouse may have been introduced in some ship from the S. American coast (though the species is at present unknown there), for the aboriginal species soon haunts the goods of man, as I noticed in the roof of a newly erected shed in a desert country south of the Plata. The Falkland Islands, though between 200 and 300 miles from the S. American coast, may in one sense be considered as intimately connected with it; for it is certain that formerly many icebergs loaded with boulders were stranded on its southern coast, and the old canoes which are occasionally now stranded, show that the currents still set from Tierra del Fuego. This fact, however, does not explain the presence of the *Canis antarcticus* on the Falkland Islands, unless we suppose that it formerly lived on the mainland and became extinct there, whilst it survived on these islands, to which it was borne (as happens with its northern congener, the common wolf) on an iceberg, but this fact removes the anomaly of an island, in appearance effectually separated

[1] See *Origin*, Ed i. p. 393, vi. p. 547. The discussion is much fuller in the present Essay.

from other land, having its own species of quadruped, and makes the case like that of Java and Sumatra, each having their own rhinoceros.

Before summing up all the facts given in this section on the present condition of organic beings, and endeavouring to see how far they admit of explanation, it will be convenient to state all such facts in the past geographical distribution of extinct beings as seem anyway to concern the theory of descent.

Section Second.

Geographical distribution of extinct organisms.

I have stated that if the land of the entire world be divided into (we will say) three sections, according to the amount of difference of the terrestrial mammifers inhabiting them, we shall have three unequal divisions of (1st) Australia and its dependent islands, (2nd) South America, (3rd) Europe, Asia and Africa. If we now look to the mammifers which inhabited these three divisions during the later Tertiary periods, we shall find them almost as distinct as at the present day, and intimately related in each division to the existing forms in that division[1]. This is wonderfully the case with the several fossil Marsupial genera in the caverns of New South Wales and even more wonderfully so in South America, where we have the same peculiar group of monkeys, of a guanaco-like animal, of many rodents, of the Marsupial Didelphys, of Armadilloes and other Edentata. This last family is at present very characteristic of S. America, and in a late Tertiary epoch it was even more so, as is shown by the numerous enormous animals of the Megatheroid family, some

[1] See *Origin*, Ed. i. p. 339, vi. p. 485.

of which were protected by an osseous armour like that, but on a gigantic scale, of the recent Armadillo. Lastly, over Europe the remains of the several deer, oxen, bears, foxes, beavers, field-mice, show a relation to the present inhabitants of this region; and the contemporaneous remains of the elephant, rhinoceros, hippopotamus, hyæna, show a relation with the grand Africo-Asiatic division of the world. In Asia the fossil mammifers of the Himalaya (though mingled with forms long extinct in Europe) are equally related to the existing forms of the Africo-Asiatic division; but especially to those of India itself. As the gigantic and now extinct quadrupeds of Europe have naturally excited more attention than the other and smaller remains, the relation between the past and the present mammiferous inhabitants of Europe has not been sufficiently attended to. But in fact the mammifers of Europe are at present nearly as much Africo-Asiatic as they were formerly when Europe had its elephants and rhinoceroses, etc.: Europe neither now nor then possessed peculiar groups as does Australia and S. America. The extinction of certain peculiar forms in one quarter does not make the remaining mammifers of that quarter less related to its own great division of the world: though Tierra del Fuego possesses only a fox, three rodents, and the guanaco, no one (as these all belong to S. American types, but not to the most characteristic forms) would doubt for one minute ⟨as to⟩ classifying this district with S. America; and if fossil Edentata, Marsupials and monkeys were to be found in Tierra del Fuego, it would not make this district more truly S. American than it now is. So it is with Europe[1], and so far as

[1] In the *Origin*, Ed. i. p. 339, vi. p. 485, which corresponds to this part of the present Essay, the author does not make a separate section for such cases as the occurrence of fossil Marsupials in Europe (*Origin*, Ed. i. p. 340, vi. p. 486) as he does in the present Essay; see the section on *Changes in geographical distribution*, p. 177.

is known with Asia, for the lately past and present mammifers all belong to the Africo-Asiatic division of the world. In every case, I may add, the forms which a country has is of more importance in geographical arrangement than what it has not.

We find some evidence of the same general fact in a relation between the recent and the Tertiary sea-shells, in the different main divisions of the marine world.

This general and most remarkable relation between the lately past and present mammiferous inhabitants of the three main divisions of the world is precisely the same kind of fact as the relation between the different species of the several sub-regions of any one of the main divisions. As we usually associate great physical changes with the total extinction of one series of beings, and its succession by another series, this identity of relation between the past and the present races of beings in the same quarters of the globe is more striking than the same relation between existing beings in different sub-regions: but in truth we have no reason for supposing that a change in the conditions has in any of these cases supervened, greater than that now existing between the temperate and tropical, or between the highlands and lowlands of the same main divisions, now tenanted by related beings. Finally, then, we clearly see that in each main division of the world the same relation holds good between its inhabitants in time as over space[1].

[1] "We can understand how it is that all the forms of life, ancient and recent, make together one grand system; for all are connected by generation." *Origin*, Ed. i. p. 344, vi. p. 491.

Changes in geographical distribution.

If, however, we look closer, we shall find that even Australia, in possessing a terrestrial Pachyderm, was so far less distinct from the rest of the world than it now is; so was S. America in possessing the Mastodon, horse, [hyæna,][1] and antelope. N. America, as I have remarked, is now, in its mammifers, in some respects neutral ground between S. America and the great Africo-Asiatic division; formerly, in possessing the horse, Mastodon and three Megatheroid animals, it was more nearly related to S. America; but in the horse and Mastodon, and likewise in having the elephant, oxen, sheep, and pigs, it was as much, if not more, related to the Africo-Asiatic division. Again, northern India was much more closely related (in having the giraffe, hippopotamus, and certain musk-deer) to southern Africa than it now is; for southern and eastern Africa deserve, if we divide the world into five parts, to make one division by itself. Turning to the dawn of the Tertiary period, we must, from our ignorance of other portions of the world, confine ourselves to Europe; and at that period, in the presence of Marsupials[2] and Edentata, we behold an *entire* blending of those mammiferous forms which now eminently characterise Australia and S. America[3].

If we now look at the distribution of sea-shells, we find the same changes in distribution. The Red Sea and the Mediterranean were more nearly related in these shells than they now are. In different parts of Europe, on the other hand, during the

[1] The word *hyæna* is erased. There appear to be no fossil Hyænidæ in S. America.

[2] See note 1, p. 175, also *Origin*, Ed. i. p. 340, vi. p. 486.

[3] ⟨Note by the author.⟩ And see Eocene European mammals in N. America.

Miocene period, the sea-shells seem to have been more different than at present. In[1] the Tertiary period, according to Lyell, the shells of N. America and Europe were less related than at present, and during the Cretaceous still less like; whereas, during this same Cretaceous period, the shells of India and Europe were more like than at present. But going further back to the Carbonaceous period, in N. America and Europe, the productions were much more like than they now are[2]. These facts harmonise with the conclusions drawn from the present distribution of organic beings, for we have seen, that from species being created in different points or areas, the formation of a barrier would cause or make two distinct geographical areas; and the destruction of a barrier would permit their diffusion[3]. And as long-continued geological changes must both destroy and make barriers, we might expect, the further we looked backwards, the more changed should we find the present distribution. This conclusion is worthy of attention; because, finding in widely different parts of the same main division of the world, and in volcanic islands near them, groups of distinct, but related, species;—and finding that a singularly analogous relation holds good with respect to the beings of past times, when none of the present species were living, a person might be tempted to believe in some mystical relation between certain areas of the world, and the production of certain organic forms; but we now see that such an assumption would have to be complicated by the admission that such a relation, though holding good for long revolutions of years, is not truly persistent.

I will only add one more observation to this

[1] ⟨Note by the author.⟩ All this requires much verification.
[2] This point seems to be less insisted on in the *Origin*.
[3] *Origin*, Ed. i. p. 356, vi. p. 504.

section. Geologists finding in the most remote period with which we are acquainted, namely in the Silurian period, that the shells and other marine productions[1] in North and South America, in Europe, Southern Africa, and Western Asia, are much more similar than they now are at these distant points, appear to have imagined that in these ancient times the laws of geographical distribution were quite different than what they now are: but we have only to suppose that great continents were extended east and west, and thus did not divide the inhabitants of the temperate and tropical seas, as the continents now do; and it would then become probable that the inhabitants of the seas would be much more similar than they now are. In the immense space of ocean extending from the east coast of Africa to the eastern islands of the Pacific, which space is connected either by lines of tropical coast or by islands not very distant from each other, we know (Cuming) that many shells, perhaps even as many as 200, are common to the Zanzibar coast, the Philippines, and the eastern islands of the Low or Dangerous Archipelago in the Pacific. This space equals that from the Arctic to the Antarctic pole! Pass over the space of quite open ocean, from the Dangerous Archipelago to the west coast of S. America, and every shell is different: pass over the narrow space of S. America, to its eastern shores, and again every shell is different! Many fish, I may add, are also common to the Pacific and Indian Oceans.

[1] (Note by the author.) D'Orbigny shows that this is not so.

Summary on the distribution of living and extinct organic beings.

Let us sum up the several facts now given with respect to the past and present geographical distribution of organic beings. In a previous chapter it was shown that species are not exterminated by universal catastrophes, and that they are slowly produced: we have also seen that each species is probably only once produced, on one point or area once in time; and that each diffuses itself, as far as barriers and its conditions of life permit. If we look at any one main division of the land, we find in the different parts, whether exposed to different conditions or to the same conditions, many groups of species wholly or nearly distinct as species, nevertheless intimately related. We find the inhabitants of islands, though distinct as species, similarly related to the inhabitants of the nearest continent; we find in some cases, that even the different islands of one such group are inhabited by species distinct, though intimately related to one another and to those of the nearest continent:—thus typifying the distribution of organic beings over the whole world. We find the floras of distant mountain-summits either very similar (which seems to admit, as shown, of a simple explanation) or very distinct but related to the floras of the surrounding region; and hence, in this latter case, the floras of two mountain-summits, although exposed to closely similar conditions, will be very different. On the mountain-summits of islands, characterised by peculiar faunas and floras, the plants are often eminently peculiar. The dissimilarity of the organic beings inhabiting nearly similar countries is best seen by comparing the main divisions of the world; in each of which some districts may be found very similarly exposed, yet the inhabitants are wholly unlike;—far more unlike than those in very dissimilar districts in the same main division. We see this strikingly in comparing two volcanic archipelagoes, with nearly the same climate, but situated not very far from two different continents; in which case their inhabitants are totally unlike. In the different main divisions of the world, the amount of difference between the organisms, even in the same class, is widely different, each main division having only the species distinct in some families, in other families having the genera distinct. The distribution of aquatic organisms is very different from that of the terrestrial organisms; and necessarily so, from the barriers to their progress being quite unlike. The nature of the conditions in an isolated district will not explain the number of species inhabiting it; nor the absence of one class or the presence of another class. We find that terrestrial mammifers are not present on islands far removed from other land. We see in two regions, that the species though distinct are more or less related, according to the greater or less *possibility* of the transportal in past and present times of species from one to the other region; although we can hardly admit that all the species in such cases have been transported from the first to the second region, and since have become extinct in the first: we see this law in the presence of the fox on the Falkland Islands; in the European character of some of the plants of Tierra del Fuego; in the Indo-Asiatic character of the plants of the Pacific; and in the circumstance of those genera which range widest having many species with wide ranges; and those genera with restricted ranges having species with restricted ranges. Finally, we find in each of the main divisions of the land, and probably of the sea, that the existing organisms are related to those lately extinct.

Looking further backwards we see that the past geographical distribution of organic beings was different from the present; and indeed, considering that geology shows that all our land was once under water, and that where water now extends land is forming, the reverse could hardly have been possible.

Now these several facts, though evidently all more or less connected together, must by the creationist (though the geologist may explain some of the anomalies) be considered as so many ultimate facts. He can only say, that it so pleased the Creator that the organic beings of the plains, deserts, mountains, tropical and temperature forests, of S. America, should all have some affinity together; that the inhabitants of the Galapagos Archipelago should be related to those of Chile; and that some of the species on the similarly constituted islands of this archipelago, though most closely related, should be distinct; that all its inhabitants should be totally unlike those of the similarly volcanic and arid Cape de Verde and Canary Islands; that the plants on the summit of Teneriffe should be eminently peculiar; that the diversified island of New Zealand should have not many plants, and not one, or only one, mammifer; that the mammifers of S. America, Australia and Europe should be clearly related to their ancient and exterminated prototypes; and so on with other facts. But it is absolutely opposed to every analogy, drawn from the laws imposed by the Creator on inorganic matter, that facts, when connected, should be considered as ultimate and not the direct consequences of more general laws.

SECTION THIRD.

An attempt to explain the foregoing laws of geographical distribution, on the theory of allied species having a common descent.

First let us recall the circumstances most favourable for variation under domestication, as given in the first chapter—viz. 1st, a change, or repeated changes, in the conditions to which the organism has been exposed, continued through several seminal (*i.e.* not by buds or divisions) generations: 2nd, steady selection of the slight varieties thus generated with a fixed end in view: 3rd, isolation as perfect as possible of such selected varieties; that is, the preventing their crossing with other forms; this latter condition applies to all terrestrial animals, to most if not all plants and perhaps even to most (or all) aquatic organisms. It will be convenient here to show the advantage of isolation in the formation of a new breed, by comparing the progress of two persons (to neither of whom let time be of any consequence) endeavouring to select and form some very peculiar new breed. Let one of these persons work on the vast herds of cattle in the plains of La Plata[1], and the other on a small stock of 20 or 30 animals in an island. The latter might have to wait centuries (by the hypothesis of no importance)[2] before he obtained a "sport" approaching to what he wanted; but when he did and saved the greater number of its offspring and their offspring again, he might hope that his whole little stock would be in some degree affected, so that by continued selection he might

[1] This instance occurs in the Essay of 1842, p. 32, but not in the *Origin*; though the importance of isolation is discussed (*Origin*, Ed. i. p. 104, vi. p. 127).

[2] The meaning of the words within parenthesis is obscure.

gain his end. But on the Pampas, though the man might get his first approach to his desired form sooner, how hopeless would it be to attempt, by saving its offspring amongst so many of the common kind, to affect the whole herd: the effect of this one peculiar "sport[1]" would be quite lost before he could obtain a second original sport of the same kind. If, however, he could separate a small number of cattle, including the offspring of the desirable "sport," he might hope, like the man on the island, to effect his end. If there be organic beings of which two individuals *never* unite, then simple selection whether on a continent or island would be equally serviceable to make a new and desirable breed; and this new breed might be made in surprisingly few years from the great and geometrical powers of propagation to beat out the old breed; as has happened (notwithstanding crossing) where good breeds of dogs and pigs have been introduced into a limited country,—for instance, into the islands of the Pacific.

Let us now take the simplest natural case of an islet upheaved by the volcanic or subterranean forces in a deep sea, at such a distance from other land that only a few organic beings at rare intervals were transported to it, whether borne by the sea[2] (like the seeds of plants to coral-reefs), or by hurricanes, or by floods, or on rafts, or in roots of large trees, or the germs of one plant or animal attached to or in the stomach of some other animal, or by the intervention (in most cases the most probable means) of other islands since sunk or destroyed. It may be remarked that when one part of the earth's crust is raised it is probably the

[1] It is unusual to find the author speaking of the selection of *sports* rather than small variations.

[2] This brief discussion is represented in the *Origin*, Ed. i. by a much fuller one (pp. 356, 383, vi. pp. 504, 535). See, however, the section in the present Essay, p. 168.

general rule that another part sinks. Let this island go on slowly, century after century, rising foot by foot; and in the course of time we shall have instead ⟨of⟩ a small mass of rock[1], lowland and highland, moist woods and dry sandy spots, various soils, marshes, streams and pools: under water on the sea shore, instead of a rocky steeply shelving coast we shall have in some parts bays with mud, sandy beaches and rocky shoals. The formation of the island by itself must often slightly affect the surrounding climate. It is impossible that the first few transported organisms could be perfectly adapted to all these stations; and it will be a chance if those successively transported will be so adapted. The greater number would probably come from the lowlands of the nearest country; and not even all these would be perfectly adapted to the new islet whilst it continued low and exposed to coast influences. Moreover, as it is certain that all organisms are nearly as much adapted in their structure to the other inhabitants of their country as they are to its physical conditions, so the mere fact that a *few* beings (and these taken in great degree by chance) were in the first case transported to the islet, would in itself greatly modify their conditions[2]. As the island continued rising we might also expect an occasional new visitant; and I repeat that even one new being must often affect beyond our calculation by occupying the room and taking part of the subsistence of another (and this again from another and so on), several or many other organisms. Now as the first transported and any occasional successive visitants spread or tended to spread over the growing island, they would undoubtedly be exposed through several generations to new and varying conditions: it might also easily happen that some of

[1] On the formation of new stations, see *Origin*, Ed. i. p. 292, vi. p. 429.
[2] *Origin*, Ed. i. pp. 390, 400, vi. pp. 543, 554.

the species *on an average* might obtain an increase of food, or food of a more nourishing quality[1]. According then to every analogy with what we have seen takes place in every country, with nearly every organic being under domestication, we might expect that some of the inhabitants of the island would "sport," or have their organization rendered in some degree plastic. As the number of the inhabitants are supposed to be few and as all these cannot be so well adapted to their new and varying conditions as they were in their native country and habitat, we cannot believe that every place or office in the economy of the island would be as well filled as on a continent where the number of aboriginal species is far greater and where they consequently hold a more strictly limited place. We might therefore expect on our island that although very many slight variations were of no use to the plastic individuals, yet that occasionally in the course of a century an individual might be born[2] of which the structure or constitution in some slight degree would allow it better to fill up some office in the insular economy and to struggle against other species. If such were the case the individual and its offspring would have a better *chance* of surviving and of beating out its parent form; and if (as is probable) it and its offspring crossed with the unvaried parent form, yet the number of the individuals being not very great, there would be a chance of the new and more serviceable form being nevertheless in some slight degree preserved. The struggle for existence would go on annually selecting such individuals until a new race or species was formed. Either few or all the first visitants to the island might become modified, ac-

[1] In the MS. *some of the species...nourishing quality* is doubtfully erased. It seems clear that he doubted whether such a problematical supply of food would be likely to cause variation.

[2] At this time the author clearly put more faith in the importance of sport-like variation than in later years.

cording as the physical conditions of the island and those resulting from the kind and number of other transported species were different from those of the parent country—according to the difficulties offered to fresh immigration—and according to the length of time since the first inhabitants were introduced. It is obvious that whatever was the country, generally the nearest from which the first tenants were transported, they would show an affinity, even if all had become modified, to the natives of that country and even if the inhabitants of the same source (?) had been modified. On this view we can at once understand the cause and meaning of the affinity of the fauna and flora of the Galapagos Islands with that of the coast of S. America; and consequently why the inhabitants of these islands show not the smallest affinity with those inhabiting other volcanic islands, with a very similar climate and soil, near the coast of Africa[1].

To return once again to our island, if by the continued action of the subterranean forces other neighbouring islands were formed, these would generally be stocked by the inhabitants of the first island, or by a few immigrants from the neighbouring mainland; but if considerable obstacles were interposed to any communication between the terrestrial productions of these islands, and their conditions were different (perhaps only by the number of different species on each island), a form transported from one island to another might become altered in the same manner as one from the continent; and we should have several of the islands tenanted by representative races or species, as is so wonderfully the case with the different islands of the Galapagos Archipelago. As the islands become mountainous, if mountain-species were not introduced, as could rarely happen, a greater amount of variation and

[1] *Origin*, Ed. i. p. 398, vi. p. 553.

selection would be requisite to adapt the species, which originally came from the lowlands of the nearest continent, to the mountain-summits than to the lower districts of our islands. For the lowland species from the continent would have first to struggle against other species and other conditions on the coast-land of the island, and so probably become modified by the selection of its best fitted varieties, then to undergo the same process when the land had attained a moderate elevation; and then lastly when it had become Alpine. Hence we can understand why the faunas of insular mountain-summits are, as in the case of Teneriffe, eminently peculiar. Putting on one side the case of a widely extended flora being driven up the mountain-summits, during a change of climate from cold to temperate, we can see why in other cases the floras of mountain-summits (or as I have called them islands in a sea of land) should be tenanted by peculiar species, but related to those of the surrounding lowlands, as are the inhabitants of a real island in the sea to those of the nearest continent[1].

Let us now consider the effect of a change of climate or of other conditions on the inhabitants of a continent and of an isolated island without any great change of level. On a continent the chief effects would be changes in the numerical proportion of the individuals of the different species; for whether the climate became warmer or colder, drier or damper, more uniform or extreme, some species are at present adapted to its diversified districts; if for instance it became cooler, species would migrate from its more temperate parts and from its higher land; if damper, from its damper

[1] See *Origin*, Ed. i. p. 403, vi. p. 558, where the author speaks of Alpine humming birds, rodents, plants, &c. in S. America, all of strictly American forms. In the MS. the author has added between the lines "As world has been getting hotter, there has been radiation from high-lands,—old view?—curious; I presume Diluvian in origin."

regions, &c. On a small and isolated island, however, with few species, and these not adapted to much diversified conditions, such changes instead of merely increasing the number of certain species already adapted to such conditions, and decreasing the number of other species, would be apt to affect the constitutions of some of the insular species: thus if the island became damper it might well happen that there were no species living in any part of it adapted to the consequences resulting from more moisture. In this case therefore, and still more (as we have seen) during the production of new stations from the elevation of the land, an island would be a far more fertile source, as far as we can judge, of new specific forms than a continent. The new forms thus generated on an island, we might expect, would occasionally be transported by accident, or through long-continued geographical changes be enabled to emigrate and thus become slowly diffused.

But if we look to the origin of a continent; almost every geologist will admit that in most cases it will have first existed as separate islands which gradually increased in size[1]; and therefore all that which has been said concerning the probable changes of the forms tenanting a small archipelago is applicable to a continent in its early state. Furthermore, a geologist who reflects on the geological history of Europe (the only region well known) will admit that it has been many times depressed, raised and left stationary. During the sinking of a continent and the probable generally accompanying changes of climate the effect would be little, *except* on the numerical proportions and in the extinction (from the lessening of rivers, the drying of marshes

[1] See the comparison between the Malay Archipelago and the probable former state of Europe, *Origin*, Ed. i. p. 299, vi. p. 438, also *Origin*, Ed. i. p. 292, vi. p. 429.

and the conversion of high-lands into low &c.) of some or of many of the species. As soon however as the continent became divided into many isolated portions or islands, preventing free immigration from one part to another, the effect of climatic and other changes on the species would be greater. But let the now broken continent, forming isolated islands, begin to rise and new stations thus to be formed, exactly as in the first case of the upheaved volcanic islet, and we shall have equally favourable conditions for the modification of old forms, that is the formation of new races or species. Let the islands become reunited into a continent; and then the new and old forms would all spread, as far as barriers, the means of transportal, and the preoccupation of the land by other species, would permit. Some of the new species or races would probably become extinct, and some perhaps would cross and blend together. We should thus have a multitude of forms, adapted to all kinds of slightly different stations, and to diverse groups of either antagonist or food-serving species. The oftener these oscillations of level had taken place (and therefore generally the older the land) the greater the number of species ⟨which⟩ would tend to be formed. The inhabitants of a continent being thus derived in the first stage from the same original parents, and subsequently from the inhabitants of one wide area, since often broken up and reunited, all would be obviously related together and the inhabitants of the most *dissimilar* stations on the same continent would be more closely allied than the inhabitants of two very *similar* stations on two of the main divisions of the world[1].

I need hardly point out that we now can ob-

[1] *Origin*, Ed. i. p. 349, vi. p. 496. The arrangement of the argument in the present Essay leads to repetition of statements made in the earlier part of the book: in the *Origin* this is avoided.

viously see why the number of species in two districts, independently of the number of stations in such districts, should be in some cases as widely different as in New Zealand and the Cape of Good Hope[1]. We can see, knowing the difficulty in the transport of terrestrial mammals, why islands far from mainlands do not possess them[2]; we see the general reason, namely accidental transport (though not the precise reason), why certain islands should, and others should not, possess members of the class of reptiles. We can see why an ancient channel of communication between two distant points, as the Cordillera probably was between southern Chile and the United States during the former cold periods; and icebergs between the Falkland Islands and Tierra del Fuego; and gales, at a former or present time, between the Asiatic shores of the Pacific and eastern islands in this ocean; is connected with (or we may now say causes) an affinity between the species, though distinct, in two such districts. We can see how the better chance of diffusion, from several of the species of any genus having wide ranges in their own countries, explains the presence of other species of the same genus in other countries[3]; and on the other hand, of species of restricted powers of ranging, forming genera with restricted ranges.

As every one would be surprised if two exactly similar but peculiar varieties[4] of any species were raised by man by long continued selection, in two different countries, or at two very different periods, so we ought not to expect that an exactly similar form would be produced from the modification of an old one in two distinct countries or at two distinct

[1] *Origin*, Ed. i. p. 389, vi. p. 542.
[2] *Origin*, Ed. i. p. 393, vi. p. 547.
[3] *Origin*, Ed. i. pp. 350, 404, vi. pp. 498 559.
[4] *Origin*, Ed. i. p. 352, vi. p. 500.

periods. For in such places and times they would probably be exposed to somewhat different climates and almost certainly to different associates. Hence we can see why each species appears to have been produced singly, in space and in time. I need hardly remark that, according to this theory of descent, there is no necessity of modification in a species, when it reaches a new and isolated country. If it be able to survive and if slight variations better adapted to the new conditions are not selected, it might retain (as far as we can see) its old form for an indefinite time. As we see that some sub-varieties produced under domestication are more variable than others, so in nature, perhaps, some species and genera are more variable than others. The same precise form, however, would probably be seldom preserved through successive geological periods, or in widely and differently conditioned countries[1].

Finally, during the long periods of time and probably of oscillations of level, necessary for the formation of a continent, we may conclude (as above explained) that many forms would become extinct. These extinct forms, and those surviving (whether or not modified and changed in structure), will all be related in each continent in the same manner and degree, as are the inhabitants of any two different sub-regions in that same continent. I do not mean to say that, for instance, the present Marsupials of Australia or Edentata and rodents of S. America have descended from any one of the few fossils of the same orders which have been discovered in these countries. It is possible that, in a very few instances, this may be the case; but generally they must be considered as merely codescendants of common stocks[2]. I believe in this, from the improbability, considering the vast number of species, which (as

[1] *Origin*, Ed. i. p. 313, vi. p. 454.
[2] *Origin*, Ed. i. p. 341, vi. p. 487.

explained in the last chapter) must by our theory have existed, that the *comparatively* few fossils which have been found should chance to be the immediate and linear progenitors of those now existing. Recent as the yet discovered fossil mammifers of S. America are, who will pretend to say that very many intermediate forms may not have existed? Moreover, we shall see in the ensuing chapter that the very existence of genera and species can be explained only by a few species of each epoch leaving modified successors or new species to a future period; and the more distant that future period, the fewer will be the *linear* heirs of the former epoch. As by our theory, all mammifers must have descended from the same parent stock, so is it necessary that each land now possessing terrestrial mammifers shall at some time have been so far united to other land as to permit the passage of mammifers[1]; and it accords with this necessity, that in looking far back into the earth's history we find, first changes in the geographical distribution, and secondly a period when the mammiferous forms most distinctive of two of the present main divisions of the world were living together[2].

I think then I am justified in asserting that most of the above enumerated and often trivial points in the geographical distribution of past and present organisms (which points must be viewed by the creationists as so many ultimate facts) follow as a simple consequence of specific forms being mutable and of their being adapted by natural selection to diverse ends, conjoined with their powers of dispersal, and the geologico-geographical changes now in slow progress and which undoubtedly have taken place. This large class of facts being thus explained,

[1] *Origin*, Ed. i. p. 396, vi. p. 549.
[2] *Origin*, Ed. i. p. 340, vi. p. 486.

far more than counterbalances many separate difficulties and apparent objections in convincing my mind of the truth of this theory of common descent.

Improbability of finding fossil forms intermediate between existing species.

There is one observation of considerable importance that may be here introduced, with regard to the improbability of the chief transitional forms between any two species being found fossil. With respect to the finer shades of transition, I have before remarked that no one has any cause to expect to trace them in a fossil state, without he be bold enough to imagine that geologists at a future epoch will be able to trace from fossil bones the gradations between the Short-Horns, Herefordshire, and Alderney breeds of cattle[1]. I have attempted to show that rising islands, in process of formation, must be the best nurseries of new specific forms, and these points are the least favourable for the embedment of fossils[2]: I appeal, as evidence, to the state of the *numerous* scattered islands in the several great oceans: how rarely do any sedimentary deposits occur on them; and when present they are mere narrow fringes of no great antiquity, which the sea is generally wearing away and destroying. The cause of this lies in isolated islands being generally volcanic and rising points; and the effects of subterranean elevation is to bring up the surrounding newly-deposited strata within the destroying action of the coast-waves: the strata, deposited at greater distances, and therefore in the depths of the ocean, will be almost barren of organic remains. These

[1] *Origin*, Ed. i. p. 299, vi. p. 437.

[2] "Nature may almost be said to have guarded against the frequent discovery of her transitional or linking forms," *Origin*, Ed. i. p. 292. A similar but not identical passage occurs in *Origin*, Ed. vi. p. 428.

remarks may be generalised:—periods of subsidence will always be most favourable to an accumulation of great thicknesses of strata, and consequently to their long preservation; for without one formation be protected by successive strata, it will seldom be preserved to a distant age, owing to the enormous amount of denudation, which seems to be a general contingent of time[1]. I may refer, as evidence of this remark, to the vast amount of subsidence evident in the great pile of the European formations, from the Silurian epoch to the end of the Secondary, and perhaps to even a later period. Periods of elevation on the other hand cannot be favourable to the accumulation of strata and their preservation to distant ages, from the circumstance just alluded to, viz. of elevation tending to bring to the surface the circum-littoral strata (always abounding most in fossils) and destroying them. The bottom of tracts of deep water (little favourable, however, to life) must be excepted from this unfavourable influence of elevation. In the quite open ocean, probably no sediment[2] is accumulating, or at a rate so slow as not to preserve fossil remains, which will always be subject to disintegration. Caverns, no doubt, will be equally likely to preserve terrestrial fossils in periods of elevation and of subsidence; but whether it be owing to the enormous amount of denudation, which all land seems to have undergone, no cavern with fossil bones has been found belonging to the Secondary period[3].

Hence many more remains will be preserved to a distant age, in any region of the world, during periods of its subsidence[4], than of its elevation.

[1] *Origin*, Ed. i. p. 291, vi. p. 426.
[2] *Origin*, Ed. i. p. 288, vi. p. 422.
[3] *Origin*, Ed. i. p. 289, vi. p. 423.
[4] *Origin*, Ed. i. p. 300, vi. p. 439.

But during the subsidence of a tract of land, its inhabitants (as before shown) will from the decrease of space and of the diversity of its stations, and from the land being fully preoccupied by species fitted to diversified means of subsistence, be little liable to modification from selection, although many may, or rather must, become extinct. With respect to its circum-marine inhabitants, although during a change from a continent to a *great* archipelago, the number of stations fitted for marine beings will be increased, their means of diffusion (an important check to change of form) will be greatly improved; for a continent stretching north and south, or a quite open space of ocean, seems to be to them the only barrier. On the other hand, during the elevation of a small archipelago and its conversion into a continent, we have, whilst the number of stations are increasing, both for aquatic and terrestrial productions, and whilst these stations are not fully preoccupied by perfectly adapted species, the most favourable conditions for the selection of new specific forms; but few of them in their early transitional states will be preserved to a distant epoch. We must wait during an enormous lapse of time, until long-continued subsidence shall have taken the place in this quarter of the world of the elevatory process, for the best conditions of the embedment and the preservation of its inhabitants. Generally the great mass of the strata in every country, from having been chiefly accumulated during subsidence, will be the tomb, not of transitional forms, but of those either becoming extinct or remaining unmodified.

The state of our knowledge, and the slowness of the changes of level, do not permit us to test the truth of these remarks, by observing whether there are more transitional or "fine" (as naturalists would term them) species, on a rising and enlarging tract of land, than on an area of subsidence. Nor do I know whether there are more "fine" species on isolated volcanic islands in process of formation, than on a continent; but I may remark, that at the Galapagos Archipelago the number of forms, which according to some naturalists are true species, and according to others are mere races, is considerable: this particularly applies to the different species or races of the same genera inhabiting the different islands of this archipelago. Furthermore it may be added (as bearing on the great facts discussed in this chapter) that when naturalists confine their attention to any one country, they have comparatively little difficulty in determining what forms to call species and what to call varieties; that is, those which can or cannot be traced or shown to be probably descendants of some other form: but the difficulty increases, as species are brought from many stations, countries and islands. It was this increasing (but I believe in few cases insuperable) difficulty which seems chiefly to have urged Lamarck to the conclusion that species are mutable.

CHAPTER VII

ON THE NATURE OF THE AFFINITIES AND CLASSIFICATION OF ORGANIC BEINGS[1]

Gradual appearance and disappearance of groups.

IT has been observed from the earliest times that organic beings fall into groups[2], and these groups into others of several values, such as species into genera, and then into sub-families, into families, orders, &c. The same fact holds with those beings which no longer exist. Groups of species seem to follow the same laws in their appearance and extinction[3], as do the individuals of any one species: we have reason to believe that, first, a few species appear, that their numbers increase; and that, when tending to extinction, the numbers of the species decrease, till finally the group becomes extinct, in the same way as a species becomes extinct, by the individuals becoming rarer and rarer. Moreover, groups, like the individuals of a species, appear to become extinct at different times in different countries. The Palæotherium was extinct

[1] Ch. XIII of the *Origin*, Ed. i., Ch. XIV Ed. vi. begins with a similar statement. In the present Essay the author adds a note:—"The obviousness of the fact ⟨*i.e.* the natural grouping of organisms⟩ alone prevents it being remarkable. It is scarcely explicable by creationist: groups of aquatic, of vegetable feeders and carnivorous, &c., might resemble each other; but why as it is. So with plants,—analogical resemblance thus accounted for. Must not here enter into details." This argument is incorporated with the text in the *Origin*, Ed. i.

[2] *Origin*, Ed. i. p. 411, vi. p. 566.

[3] *Origin*, Ed. i. p. 316, vi. p. 457.

much sooner in Europe than in India: the Trigonia[1] was extinct in early ages in Europe, but now lives in the seas of Australia. As it happens that one species of a family will endure for a much longer period than another species, so we find that some whole groups, such as Mollusca, tend to retain their forms, or to remain persistent, for longer periods than other groups, for instance than the Mammalia. Groups therefore, in their appearance, extinction, and rate of change or succession, seem to follow nearly the same laws with the individuals of a species[2].

What is the Natural System?

The proper arrangement of species into groups, according to the natural system, is the object of all naturalists; but scarcely two naturalists will give the same answer to the question, What is the natural system and how are we to recognise it? The most important characters[3] it might be thought (as it was by the earliest classifiers) ought to be drawn from those parts of the structure which determine its habits and place in the economy of nature, which we may call the final end of its existence. But nothing is further from the truth than this; how much external resemblance there is between the little otter (Chironectes) of Guiana and the common otter; or again between the common swallow and the swift; and who can doubt that the means and ends of their existence are closely similar, yet how grossly wrong would be the classification, which put close to each other a Marsupial and Placental animal, and two birds with widely different skeletons. Relations, such as in the two latter cases, or as that

[1] *Origin*, Ed. i. p. 321, vi. p. 463.

[2] In the *Origin*, Ed. i. this preliminary matter is replaced (pp. 411, 412, vi. pp. 566, 567) by a discussion in which extinction is also treated, but chiefly from the point of view of the theory of divergence.

[3] *Origin*, Ed. i. p. 414, vi. p. 570.

between the whale and fishes, are denominated "analogical[1]," or are sometimes described as "relations of adaption." They are infinitely numerous and often very singular; but are of no use in the classification of the higher groups. How it comes, that certain parts of the structure, by which the habits and functions of the species are settled, are of no use in classification, whilst other parts, formed at the same time, are of the greatest, it would be difficult to say, on the theory of separate creations.

Some authors as Lamarck, Whewell &c., believe that the degree of affinity on the natural system depends on the degrees of resemblance in organs more or less physiologically important for the preservation of life. This scale of importance in the organs is admitted to be of difficult discovery. But quite independent of this, the proposition, as a general rule, must be rejected as false; though it may be partially true. For it is universally admitted that the same part or organ, which is of the highest service in classification in one group, is of very little use in another group, though in both groups, as far as we can see, the part or organ is of equal physiological importance: moreover, characters quite unimportant physiologically, such as whether the covering of the body consists of hair or feathers, whether the nostrils communicated with the mouth[2] &c., &c., are of the highest generality in classification; even colour, which is so inconstant in many species, will sometimes well characterise even a whole group of species. Lastly, the fact, that no one character is of so much importance in determining to what great group an organism belongs, as the forms through which the embryo[3] passes from the germ upwards to maturity, cannot be reconciled

[1] *Origin*, Ed. i. p. 414, vi. p. 570.
[2] These instances occur with others in the *Origin*, Ed. i. p. 416, vi. p. 572.
[3] *Origin*, Ed. i. p. 418, vi. p. 574.

with the idea that natural classification follows according to the degrees of resemblance in the parts of most physiological importance. The affinity of the common rock-barnacle with the Crustaceans can hardly be perceived in more than a single character in its mature state, but whilst young, locomotive, and furnished with eyes, its affinity cannot be mistaken[1]. The cause of the greater value of characters, drawn from the early stages of life, can, as we shall in a succeeding chapter see, be in a considerable degree explained, on the theory of descent, although inexplicable on the views of the creationist.

Practically, naturalists seem to classify according to the resemblance of those parts or organs which in related groups are most uniform, or vary least[2]: thus the æstivation, or manner in which the petals etc. are folded over each other, is found to afford an unvarying character in most families of plants, and accordingly any difference in this respect would be sufficient to cause the rejection of a species from many families; but in the Rubiaceæ the æstivation is a varying character, and a botanist would not lay much stress on it, in deciding whether or not to class a new species in this family. But this rule is obviously so arbitrary a formula, that most naturalists seem to be convinced that something ulterior is represented by the natural system; they appear to think that we only discover by such similarities what the arrangement of the system is, not that such similarities make the system. We can only thus understand Linnæus'[3] well-known saying, that the characters do not make the genus; but that the genus gives the characters: for a classification, independent of characters, is here presupposed.

[1] *Origin*, Ed. i. pp. 419, 440, vi. pp. 575, 606.
[2] *Origin*, Ed. i. pp. 418, 425, vi. pp. 574, 581.
[3] *Origin*, Ed. i. p. 413, vi. p. 569.

Hence many naturalists have said that the natural system reveals the plan of the Creator: but without it be specified whether order in time or place, or what else is meant by the plan of the Creator, such expressions appear to me to leave the question exactly where it was.

Some naturalists consider that the geographical position[1] of a species may enter into the consideration of the group into which it should be placed; and most naturalists (either tacitly or openly) give value to the different groups, not solely by their relative differences in structure, but by the number of forms included in them. Thus a genus containing a few species might be, and has often been, raised into a family on the discovery of several other species. Many natural families are retained, although most closely related to other families, from including a great number of closely similar species. The more logical naturalist would perhaps, if he could, reject these two contingents in classification. From these circumstances, and especially from the undefined objects and criterions of the natural system, the number of divisions, such as genera, sub-families, families, &c., &c., has been quite arbitrary[2]; without the clearest definition, how can it be possible to decide whether two groups of species are of equal value, and of what value? whether they should both be called genera or families; or whether one should be a genus, and the other a family[3]?

[1] *Origin*, Ed. i. pp. 419, 427, vi. pp. 575, 582.

[2] This is discussed from the point of view of divergence in the *Origin*, Ed. i. pp. 420, 421, vi. pp. 576, 577.

[3] ⟨Footnote by the author.⟩ I discuss this because if Quinarism true, I false. ⟨The Quinary System is set forth in W. S. Macleay's *Horæ Entomologicæ*, 1821.⟩

On the kind of relation between distinct groups.

I have only one other remark on the affinities of organic beings; that is, when two quite distinct groups approach each other, the approach is *generally* generic[1] and not special; I can explain this most easily by an example: of all Rodents the Bizcacha, by certain peculiarities in its reproductive system, approaches nearest to the Marsupials; of all Marsupials the Phascolomys, on the other hand, appears to approach in the form of its teeth and intestines nearest to the Rodents; but there is no special relation between these two genera[2]; the Bizcacha is no nearer related to the Phascolomys than to any other Marsupial in the points in which it approaches this division; nor again is the Phascolomys, in the points of structure in which it approaches the Rodents, any nearer related to the Bizcacha than to any other Rodent. Other examples might have been chosen, but I have given (from Waterhouse) this example as it illustrates another point, namely, the difficulty of determining what are analogical or adaptive and what real affinities; it seems that the teeth of the Phascolomys though *appearing closely* to resemble those of a Rodent are found to be built on the Marsupial type; and it is thought that these teeth and consequently the intestines may have been adapted to the peculiar life of this animal and therefore may not show any real relation. The structure in the Bizcacha that connects it with the Marsupials does not seem a peculiarity related to its manner of life, and I imagine that no one would doubt that this shows a real affinity, though not more with any one Marsupial

[1] In the corresponding passage in the *Origin*, Ed. i. p. 430, vi. p. 591, the term *general* is used in place of *generic*, and seems a better expression. In the margin the author gives Waterhouse as his authority.

[2] *Origin*, Ed. i. p. 430, vi. p. 591.

species than with another. The difficulty of determining what relations are real and what analogical is far from surprising when no one pretends to define the meaning of the term relation or the ulterior object of all classification. We shall immediately see on the theory of descent how it comes that there should be "real" and "analogical" affinities; and why the former alone should be of value in classification—difficulties which it would be I believe impossible to explain on the ordinary theory of separate creations.

Classification of Races or Varieties.

Let us now for a few moments turn to the classification of the generally acknowledged varieties and subdivisions of our domestic beings[1]; we shall find them systematically arranged in groups of higher and higher value. De Candolle has treated the varieties of the cabbage exactly as he would have done a natural family with various divisions and subdivisions. In dogs again we have one main division which may be called the *family* of hounds; of these, there are several (we will call them) *genera*, such as blood-hounds, fox-hounds, and harriers; and of each of these we have different *species*, as the blood-hound of Cuba and that of England; and of the latter again we have breeds truly producing their own kind, which may be called races or varieties. Here we see a classification practically used which typifies on a lesser scale that which holds good in nature. But amongst true species in the natural system and amongst domestic races the number of divisions or groups, instituted between those most alike and those most unlike, seems to be quite

[1] In a corresponding passage in the *Origin*, Ed. i. p. 423, vi. p. 579, the author makes use of his knowledge of pigeons. The pseudo-genera among dogs are discussed in *Var. under Dom.*, Ed. ii. vol. I. p. 38.

arbitrary. The number of the forms in both cases seems practically, whether or not it ought theoretically, to influence the denomination of groups including them. In both, geographical distribution has sometimes been used as an aid to classification[1]; amongst varieties, I may instance, the cattle of India or the sheep of Siberia, which from possessing some characters in common permit a classification of Indian and European cattle, or Siberian and European sheep. Amongst domestic varieties we have even something very like the relations of "analogy" or "adaptation[2]"; thus the common and Swedish turnip are both artificial varieties which strikingly resemble each other, and they fill nearly the same end in the economy of the farm-yard; but although the swede so much more resembles a turnip than its presumed parent the field cabbage, no one thinks of putting it out of the cabbages into the turnips. Thus the greyhound and racehorse, having been selected and trained for extreme fleetness for short distances, present an analogical resemblance of the same kind, but less striking as that between the little otter (Marsupial) of Guiana and the common otter; though these two otters are really less related than ⟨are⟩ the horse and dog. We are even cautioned by authors treating on varieties, to follow the *natural* in contradistinction of an artificial system and not, for instance, to class two varieties of the pine-apple[3] near each other, because their fruits accidentally resemble each other closely (though the fruit may be called the *final end* of this plant in the economy of its world, the hot-house), but to judge from the general resemblance of the entire plants. Lastly, varieties often become extinct; sometimes from unexplained causes, some-

[1] *Origin*, Ed. i. pp. 419, 427, vi. pp. 575, 582.
[2] *Origin*, Ed. i. pp. 423, 427, vi. pp. 579, 583.
[3] *Origin*, Ed. i. p. 423, vi. p. 579.

times from accident, but more often from the production of more useful varieties, and the less useful ones being destroyed or bred out.

I think it cannot be doubted that the main cause of all the varieties which have descended from the aboriginal dog or dogs, or from the aboriginal wild cabbage, not being equally like or unlike—but on the contrary, obviously falling into groups and sub-groups—must in chief part be attributed to different degrees of true relationship; for instance, that the different kinds of blood-hound have descended from one stock, whilst the harriers have descended from another stock, and that both these have descended from a different stock from that which has been the parent of the several kinds of grey-hound. We often hear of a florist having some choice variety and breeding from it a whole group of sub-varieties more or less characterised by the peculiarities of the parent. The case of the peach and nectarine, each with their many varieties, might have been introduced. No doubt the relationship of our different domestic breeds has been obscured in an extreme degree by their crossing; and likewise from the slight difference between many breeds it has probably often happened that a "sport" from one breed has less closely resembled its parent breed than some other breed, and has therefore been classed with the latter. Moreover the effects of a similar climate[1] may in some cases have more than counterbalanced the similarity, consequent on a common descent, though I should think the similarity of the breeds of cattle of India or sheep of Siberia was far more probably due to the community of their descent than to the effects of climate on animals descended from different stocks.

Notwithstanding these great sources of difficulty,

[1] A general statement of the influence of conditions on variation occurs in the *Origin*, Ed. i. pp. 131–3, vi. pp. 164–5.

I apprehend every one would admit, that if it were possible, a genealogical classification of our domestic varieties would be the most satisfactory one; and as far as varieties were concerned would be the natural system: in some cases it has been followed. In attempting to follow out this object a person would have to class a variety, whose parentage he did not know, by its external characters; but he would have a distinct ulterior object in view, namely, its descent in the same manner as a regular systematist seems also to have an ulterior but undefined end in all his classifications. Like the regular systematist he would not care whether his characters were drawn from more or less important organs as long as he found in the tribe which he was examining that the characters from such parts were persistent; thus amongst cattle he does value a character drawn from the form of the horns more than from the proportions of the limbs and whole body, for he finds that the shape of the horns is to a considerable degree persistent amongst cattle[1], whilst the bones of the limbs and body vary. No doubt as a frequent rule the more important the organ, as being less related to external influences, the less liable it is to variation; but he would expect that according to the object for which the races had been selected, parts more or less important might differ; so that characters drawn from parts generally most liable to vary, as colour, might in some instances be highly serviceable—as is the case. He would admit that general resemblances scarcely definable by language might sometimes serve to allocate a species by its nearest relation. He would be able to assign a clear reason why the close similarity of the fruit in two varieties of pine-apple, and of the so-called root in the common and Swedish turnips, and why the

[1] *Origin*, Ed. i. p. 423, vi. p. 579. In the margin Marshall is given as the authority.

similar gracefulness of form in the greyhound and racehorse, are characters of little value in classification; namely, because they are the result, not of community of descent, but either of selection for a common end, or of the effects of similar external conditions.

Classification of "races" and species similar.

Thus seeing that both the classifiers of species and of varieties[1] work by the same means, make similar distinctions in the value of the characters, and meet with similar difficulties, and that both seem to have in their classification an ulterior object in view; I cannot avoid strongly suspecting that the same cause, which has made amongst our domestic varieties groups and sub-groups, has made similar groups (but of higher values) amongst species; and that this cause is the greater or less propinquity of actual descent. The simple fact of species, both those long since extinct and those now living, being divisible into genera, families, orders &c.—divisions analogous to those into which varieties are divisible—is otherwise an inexplicable fact, and only not remarkable from its familiarity.

Origin of genera and families.

Let us suppose[2] for example that a species spreads and arrives at six or more different regions, or being already diffused over one wide area, let this area be divided into six distinct regions, exposed to different conditions, and with stations slightly different, not fully occupied with other species, so

[1] *Origin*, Ed. i. p. 423, vi. p. 579.

[2] The discussion here following corresponds more or less to the *Origin*, Ed. i. pp. 411, 412, vi. pp. 566, 567; although the doctrine of divergence is not mentioned in this Essay (as it is in the *Origin*) yet the present section seems to me a distinct approximation to it.

that six different races or species were formed by selection, each best fitted to its new habits and station. I must remark that in every case, if a species becomes modified in any one sub-region, it is probable that it will become modified in some other of the sub-regions over which it is diffused, for its organization is shown to be capable of being rendered plastic; its diffusion proves that it is able to struggle with the other inhabitants of the several sub-regions; and as the organic beings of every great region are in some degree allied, and as even the physical conditions are often in some respects alike, we might expect that a modification in structure, which gave our species some advantage over antagonist species in one sub-region, would be followed by other modifications in other of the sub-regions. The races or new species supposed to be formed would be closely related to each other; and would either form a new genus or sub-genus, or would rank (probably forming a slightly different section) in the genus to which the parent species belonged. In the course of ages, and during the contingent physical changes, it is probable that some of the six new species would be destroyed; but the same advantage, whatever it may have been (whether mere tendency to vary, or some peculiarity of organization, power of mind, or means of distribution), which in the parent-species and in its six selected and changed species-offspring, caused them to prevail over other antagonist species, would generally tend to preserve some or many of them for a long period. If then, two or three of the six species were preserved, they in their turn would, during continued changes, give rise to as many small groups of species: if the parents of these small groups were closely similar, the new species would form one great genus, barely perhaps divisible into two or three sections: but if the parents were considerably unlike, their species-offspring would, from inheriting most of the peculiarities of their parent-stocks, form either two or more sub-genera or (if the course of selection tended in different ways) genera. And lastly species descending from different species of the newly formed genera would form new genera, and such genera collectively would form a family.

The extermination of species follows from changes in the external conditions, and from the increase or immigration of more favoured species: and as those species which are undergoing modification in any one great region (or indeed over the world) will very often be allied ones from (as just explained) partaking of many characters, and therefore advantages in common, so the species, whose place the new or more favoured ones are seizing, from partaking of a common inferiority (whether in any particular point of structure, or of general powers of mind, of means of distribution, of capacity for variation, &c., &c.), will be apt to be allied. Consequently species of the same genus will slowly, one after the other, *tend* to become rarer and rarer in numbers, and finally extinct; and as each last species of several allied genera fails, even the family will become extinct. There may of course be occasional exceptions to the entire destruction of any genus or family. From what has gone before, we have seen that the slow and successive formation of several new species from the same stock will make a new genus, and the slow and successive formation of several other new species from another stock will make another genus; and if these two stocks were allied, such genera will make a new family. Now, as far as our knowledge serves, it is in this slow and gradual manner that groups of species appear on, and disappear from, the face of the earth.

The manner in which, according to our theory, the arrangement of species in groups is due to partial extinction, will perhaps be rendered clearer in the following way. Let us suppose in any one great class, for instance in the Mammalia, that every species and every variety, during each successive age, had sent down one unaltered descendant (either fossil or living) to the present time; we should then have had one enormous series, including by small gradations every known mammiferous form; and consequently the existence of groups[1], or chasms in the series, which in some parts are in greater width, and in some of less, is solely due to former species, and whole groups of species, not having thus sent down descendants to the present time.

With respect to the "analogical" or "adaptive" resemblances between organic beings which are not really related[2], I will only add, that probably the isolation of different groups of species is an important element in the production of such characters: thus we can easily see, in a large increasing island, or even a continent like Australia, stocked with only certain orders of the main classes, that the conditions would be highly favourable for species from these orders to become adapted to play parts in the economy of nature, which in other countries were performed by tribes especially adapted to such parts. We can understand how it might happen that an otter-like animal might have been formed in Australia by slow selection from the more carnivorous Marsupial types; thus we can understand that curious case in the southern hemisphere, where there are no auks (but many petrels), of a petrel[3] having been modified into the

[1] The author probably intended to write "groups separated by chasms."
[2] A similar discussion occurs in the *Origin*, Ed. i. p. 427, vi. p. 582.
[3] *Puffinuria berardi*, see *Origin*, Ed. i. p. 184, vi. p. 221.

external general form so as to play the same office in nature with the auks of the northern hemisphere; although the habits and form of the petrels and auks are normally so wholly different. It follows, from our theory, that two orders must have descended from one common stock at an immensely remote epoch; and we can perceive when a species in either order, or in both, shows some affinity to the other order, why the affinity is usually generic and not particular—that is why the Bizcacha amongst Rodents, in the points in which it is related to the Marsupial, is related to the whole group[1], and not particularly to the Phascolomys, which of all Marsupialia is related most to the Rodents. For the Bizcacha is related to the present Marsupialia, only from being related to their common parent-stock; and not to any one species in particular. And generally, it may be observed in the writings of most naturalists, that when an organism is described as intermediate between two *great* groups, its relations are not to particular species of either group, but to both groups, as wholes. A little reflection will show how exceptions (as that of the Lepidosiren, a fish closely related to *particular* reptiles) might occur, namely from a few descendants of those species, which at a very early period branched out from a common parent-stock and so formed the two orders or groups, having survived, in nearly their original state, to the present time.

Finally, then, we see that all the leading facts in the affinities and classification of organic beings can be explained on the theory of the natural system being simply a genealogical one. The similarity of the principles in classifying domestic varieties and true species, both those living and extinct, is at once

[1] *Origin*, Ed. i. p. 430, vi. p. 591.

explained; the rules followed and difficulties met with being the same. The existence of genera, families, orders, &c., and their mutual relations, naturally ensues from extinction going on at all periods amongst the diverging descendants of a common stock. These terms of affinity, relations, families, adaptive characters, &c., which naturalists cannot avoid using, though metaphorically, cease being so, and are full of plain signification.

CHAPTER VIII

UNITY OF TYPE IN THE GREAT CLASSES; AND MORPHOLOGICAL STRUCTURES

Unity of Type[1].

SCARCELY anything is more wonderful or has been oftener insisted on than that the organic beings in each great class, though living in the most distant climes and at periods immensely remote, though fitted to widely different ends in the economy of nature, yet all in their internal structure evince an obvious uniformity. What, for instance, is more wonderful than that the hand to clasp, the foot or hoof to walk, the bat's wing to fly, the porpoise's fin[2] to swim, should all be built on the same plan? and that the bones in their position and number should be so similar that they can all be classed and called by the same names. Occasionally some of the bones are merely represented by an apparently useless, smooth style, or are soldered closely to other bones, but the unity of type is not by this destroyed, and hardly rendered less clear. We see in this fact some deep bond of union between the organic beings of the same great classes—to illustrate which is the object and foundation of the natural

[1] *Origin*, Ed. i. p. 434, vi. p. 595. Ch. VIII corresponds to a section of Ch. XIII in the *Origin*, Ed. i.

[2] *Origin*, Ed. i. p. 434, vi. p. 596. In the *Origin*, Ed. i. these examples occur under the heading *Morphology*; the author does not there draw much distinction between this heading and that of *Unity of Type*.

system. The perception of this bond, I may add, is the evident cause that naturalists make an ill-defined distinction between true and adaptive affinities.

Morphology.

There is another allied or rather almost identical class of facts admitted by the least visionary naturalists and included under the name of Morphology. These facts show that in an individual organic being, several of its organs consist of some other organ metamorphosed[1]: thus the sepals, petals, stamens, pistils, &c. of every plant can be shown to be metamorphosed leaves; and thus not only can the number, position and transitional states of these several organs, but likewise their monstrous changes, be most lucidly explained. It is believed that the same laws hold good with the gemmiferous vesicles of Zoophytes. In the same manner the number and position of the extraordinarily complicated jaws and palpi of Crustacea and of insects, and likewise their differences in the different groups, all become simple, on the view of these parts, or rather legs and all metamorphosed appendages, being metamorphosed legs. The skulls, again, of the Vertebrata are composed of three metamorphosed vertebræ, and thus we can see a meaning in the number and strange complication of the bony case of the brain. In this latter instance, and in that of the jaws of the Crustacea, it is only necessary to see a series taken from the different groups of each class to admit the truth of these views. It is evident that when in each species of a group its organs consist of some other part metamorphosed, that there must also be a "unity of type" in such a group. And

[1] See *Origin*, Ed. i. p. 436, vi. p. 599, where the parts of the flower, the jaws and palpi of Crustaceans and the vertebrate skull are given as examples.

in the cases as that above given in which the foot, hand, wing and paddle are said to be constructed on a uniform type, if we could perceive in such parts or organs traces of an apparent change from some other use or function, we should strictly include such parts or organs in the department of morphology: thus if we could trace in the limbs of the Vertebrata, as we can in their ribs, traces of an apparent change from being processes of the vertebræ, it would be said that in each species of the Vertebrata the limbs were "metamorphosed spinal processes," and that in all the species throughout the class the limbs displayed a "unity of type[1]."

These wonderful parts of the hoof, foot, hand, wing, paddle, both in living and extinct animals, being all constructed on the same framework, and again of the petals, stamina, germens, &c. being metamorphosed leaves, can by the creationist be viewed only as ultimate facts and incapable of explanation; whilst on our theory of descent these facts all necessary follow: for by this theory all the beings of any one class, say of the mammalia, are supposed to be descended from one parent-stock, and to have been altered by such slight steps as man effects by the selection of chance domestic variations. Now we can see according to this view that a foot might be selected with longer and longer bones, and wider connecting membranes, till it became a swimming organ, and so on till it became an organ by which to flap along the surface or to glide over it, and lastly to fly through the air: but in such changes there would be no tendency to alter the framework of the internal inherited structure. Parts might become lost (as the tail in dogs, or horns in cattle, or the pistils in plants), others might become united together (as in the feet of the

[1] The author here brings *Unity of Type* and *Morphology* together.

Lincolnshire breed of pigs[1], and in the stamens of many garden flowers); parts of a similar nature might become increased in number (as the vertebræ in the tails of pigs, &c., &c. and the fingers and toes in six-fingered races of men and in the Dorking fowls), but analogous differences are observed in nature and are not considered by naturalists to destroy the uniformity of the types. We can, however, conceive such changes to be carried to such length that the unity of type might be obscured and finally be undistinguishable, and the paddle of the Plesiosaurus has been advanced as an instance in which the uniformity of type can hardly be recognised[2]. If after long and gradual changes in the structure of the co-descendants from any parent stock, evidence (either from monstrosities or from a graduated series) could be still detected of the function, which certain parts or organs played in the parent stock, these parts or organs might be strictly determined by their former function with the term "metamorphosed" appended. Naturalists have used this term in the same metaphorical manner as they have been obliged to use the terms of affinity and relation; and when they affirm, for instance, that the jaws of a crab are metamorphosed legs, so that one crab has more legs and fewer jaws than another, they are far from meaning that the jaws, either during the life of the individual crab or of its progenitors, were really legs. By our theory this term assumes its literal meaning[3]; and this wonderful fact of the complex jaws of an animal

[1] The solid-hoofed pigs mentioned in *Var. under Dom.*, Ed. ii. vol. II. p. 424 are not *Lincolnshire pigs*. For other cases see Bateson, *Materials for the Study of Variation*, 1894, pp. 387–90.

[2] In the margin C. Bell is given as authority, apparently for the statement about Plesiosaurus. See *Origin*, Ed. i. p. 436, vi. p. 598, where the author speaks of the "general pattern" being obscured in "extinct gigantic sea lizards." In the same place the suctorial Entomostraca are added as examples of the difficulty of recognising the type.

[3] *Origin*, Ed. i. p. 438, vi. p. 602.

retaining numerous characters, which they would probably have retained if they had really been metamorphosed during many successive generations from true legs, is simply explained.

Embryology.

The unity of type in the great classes is shown in another and very striking manner, namely, in the stages through which the embryo passes in coming to maturity[1]. Thus, for instance, at one period of the embryo, the wings of the bat, the hand, hoof or foot of the quadruped, and the fin of the porpoise do not differ, but consist of a simple undivided bone. At a still earlier period the embryo of the fish, bird, reptile and mammal all strikingly resemble each other. Let it not be supposed this resemblance is only external; for on dissection, the arteries are found to branch out and run in a peculiar course, wholly unlike that in the full-grown mammal and bird, but much less unlike that in the full-grown fish, for they run as if to ærate blood by branchiæ[2] on the neck, of which even the slit-like orifices can be discerned. How wonderful it is that this structure should be present in the embryos of animals about to be developed into such different forms, and of which two great classes respire only in the air. Moreover, as the embryo of the mammal is matured in the parent's body, and that of the bird in an egg in the air, and that of the fish in an egg in the water, we cannot believe that this course of the arteries is related to any external conditions. In all shell-fish (Gasteropods) the embryo passes through a state analogous to that of the Pteropodous Mol-

[1] *Origin*, Ed. i. p. 439, vi. p. 604.

[2] The uselessness of the branchial arches in mammalia is insisted on in the *Origin*, Ed. i. p. 440, vi. p. 606. Also the uselessness of the spots on the young blackbird and the stripes of the lion-whelp, cases which do not occur in the present Essay.

lusca: amongst insects again, even the most different ones, as the moth, fly and beetle, the crawling larvæ are all closely analogous: amongst the Radiata, the jelly-fish in its embryonic state resembles a polype, and in a still earlier state an infusorial animalcule —as does likewise the embryo of the polype. From the part of the embryo of a mammal, at one period, resembling a fish more than its parent form; from the larvæ of all orders of insects more resembling the simpler articulate animals than their parent insects[1]; and from such other cases as the embryo of the jelly-fish resembling a polype much nearer than the perfect jelly-fish; it has often been asserted that the higher animal in each class passes through the state of a lower animal; for instance, that the mammal amongst the vertebrata passes through the state of a fish[2]: but Müller denies this, and affirms that the young mammal is at no time a fish, as does Owen assert that the embryonic jelly-fish is at no time a polype, but that mammal and fish, jelly-fish and polype pass through the same state; the mammal and jelly-fish being only further developed or changed.

As the embryo, in most cases, possesses a less complicated structure than that into which it is to be developed, it might have been thought that the resemblance of the embryo to less complicated forms in the same great class, was in some manner a necessary preparation for its higher development; but in fact the embryo, during its growth, may become less, as well as more, complicated[3]. Thus certain female Epizoic Crustaceans in their mature

[1] In the *Origin*, Ed. i. pp. 442, 448, vi. pp. 608, 614 it is pointed out that in some cases the young form resembles the adult, *e.g.* in spiders; again, that in the Aphis there is no "worm-like stage" of development.

[2] In the *Origin*, Ed. i. p. 449, vi. p. 618, the author speaks doubtfully about the recapitulation theory.

[3] This corresponds to the *Origin*, Ed. i. p. 441, vi. p. 607, where, however, the example is taken from the Cirripedes.

state have neither eyes nor any organs of locomotion; they consist of a mere sack, with a simple apparatus for digestion and procreation; and when once attached to the body of the fish, on which they prey, they never move again during their whole lives: in their embryonic condition, on the other hand, they are furnished with eyes, and with well articulated limbs, actively swim about and seek their proper object to become attached to. The larvæ, also, of some moths are as complicated and are more active than the wingless and limbless females, which never leave their pupa-case, never feed and never see the daylight.

Attempt to explain the facts of embryology.

I think considerable light can be thrown by the theory of descent on these wonderful embryological facts which are common in a greater or less degree to the whole animal kingdom, and in some manner to the vegetable kingdom: on the fact, for instance, of the arteries in the embryonic mammal, bird, reptile and fish, running and branching in the same courses and nearly in the same manner with the arteries in the full-grown fish; on the fact I may add of the high importance to systematic naturalists[1] of the characters and resemblances in the embryonic state, in ascertaining the true position in the natural system of mature organic beings. The following are the considerations which throw light on these curious points.

In the economy, we will say of a feline animal[2], the feline structure of the embryo or of the sucking kitten is of quite secondary importance to it; hence, if a feline animal varied (assuming for the time the

[1] *Origin*, Ed. i. p. 449, vi. p. 617.

[2] This corresponds to the *Origin*, Ed. i. pp. 443–4, vi. p. 610: the "feline animal" is not used to illustrate the generalisation, but is so used in the Essay of 1842, p. 42.

possibility of this) and if some place in the economy of nature favoured the selection of a longer-limbed variety, it would be quite unimportant to the production by natural selection of a long-limbed breed, whether the limbs of the embryo and kitten were elongated if they *became* so *as soon* as the animal had to provide food for itself. And if it were found after continued selection and the production of several new breeds from one parent-stock, that the successive variations had supervened, not very early in the youth or embryonic life of each breed (and we have just seen that it is quite unimportant whether it does so or not), then it obviously follows that the young or embryos of the several breeds will continue resembling each other more closely than their adult parents[1]. And again, if two of these breeds became each the parent-stock of several other breeds, forming two genera, the young and embryos of these would still retain a greater resemblance to the one original stock than when in an adult state. Therefore if it could be shown that the period of the slight successive variations does not always supervene at a very early period of life, the greater resemblance or closer unity in type of animals in the young than in the full-grown state would be explained. Before practically[2] endeavouring to discover in our domestic races whether the structure or form of the young has or has not changed in an exactly corresponding degree with the changes of full-grown animals, it will be well to show that it is at least quite *possible* for the primary germinal vesicle to be impressed with a tendency to produce some change on the growing tissues which will not be fully effected till the animal is advanced in life.

[1] *Origin*, Ed. i. p. 447, vi. p. 613.

[2] In the margin is written "Get young pigeons"; this was afterwards done, and the results are given in the *Origin*, Ed. i. p. 445, vi. p. 612.

From the following peculiarities of structure being inheritable and appearing only when the animal is full-grown—namely, general size, tallness (not consequent on the tallness of the infant), fatness either over the whole body, or local; change of colour in hair and its loss; deposition of bony matter on the legs of horses; blindness and deafness, that is changes of structure in the eye and ear; gout and consequent deposition of chalk-stones; and many other diseases[1], as of the heart and brain, &c., &c.; from all such tendencies being I repeat inheritable, we clearly see that the germinal vesicle is impressed with some power which is wonderfully preserved during the production of infinitely numerous cells in the ever changing tissues, till the part ultimately to be affected is formed and the time of life arrived at. We see this clearly when we select cattle with any peculiarity of their horns, or poultry with any peculiarity of their second plumage, for such peculiarities cannot of course reappear till the animal is mature. Hence, it is certainly *possible* that the germinal vesicle may be impressed with a tendency to produce a long-limbed animal, the full proportional length of whose limbs shall appear only when the animal is mature[2].

In several of the cases just enumerated we know that the first cause of the peculiarity, when *not* inherited, lies in the conditions to which the animal is exposed during mature life, thus to a certain extent general size and fatness, lameness in horses and in a lesser degree blindness, gout and some other diseases are certainly in some degree caused

[1] In the *Origin*, Ed. i. the corresponding passages are at pp. 8, 13, 443, vi. pp. 8, 15, 610. In the *Origin*, Ed. i. I have not found a passage so striking as that which occurs a few lines lower "that the germinal vesicle is impressed with some power which is wonderfully preserved, &c." In the *Origin* this *preservation* is rather taken for granted.

[2] ⟨In the margin is written⟩ Aborted organs show, perhaps, something about period ⟨at⟩ which changes supervene in embryo.

and accelerated by the habits of life, and these peculiarities when transmitted to the offspring of the affected person reappear at a nearly corresponding time of life. In medical works it is asserted generally that at whatever period an hereditary disease appears in the parent, it tends to reappear in the offspring at the same period. Again, we find that early maturity, the season of reproduction and longevity are transmitted to corresponding periods of life. Dr Holland has insisted much on children of the same family exhibiting certain diseases in similar and peculiar manners; my father has known three brothers[1] die in very old age in a *singular* comatose state; now to make these latter cases strictly bear, the children of such families ought similarly to suffer at corresponding times of life; this is probably not the case, but such facts show that a tendency in a disease to appear at particular stages of life can be transmitted through the germinal vesicle to different individuals of the same family. It is then certainly possible that diseases affecting widely different periods of life can be transmitted. So little attention is paid to very young domestic animals that I do not know whether any case is on record of selected peculiarities in young animals, for instance, in the first plumage of birds, being transmitted to their young. If, however, we turn to silk-worms[2], we find that the caterpillars and coccoons (which must correspond to a *very early* period of the embryonic life of mammalia) vary, and that these varieties reappear in the offspring caterpillars and coccoons.

I think these facts are sufficient to render it probable that at whatever period of life any peculiarity (capable of being inherited) appears, whether caused by the action of external influences

[1] See p. 42, note.
[2] The evidence is given in *Var. under Dom.*, I. p. 316.

during mature life, or from an affection of the primary germinal vesicle, it *tends* to reappear in the offspring at the corresponding period of life[1]. Hence (I may add) whatever effect training, that is the full employment or action of every newly selected slight variation, has in fully developing and increasing such variation, would only show itself in mature age, corresponding to the period of training; in the second chapter I showed that there was in this respect a marked difference in natural and artificial selection, man not regularly exercising or adapting his varieties to new ends, whereas selection by nature presupposes such exercise and adaptation in each selected and changed part. The foregoing facts show and presuppose that slight variations occur at various periods of life *after birth*; the facts of monstrosity, on the other hand, show that many changes take place before birth, for instance, all such cases as extra fingers, hare-lip and all sudden and great alterations in structure; and these when inherited reappear during the embryonic period in the offspring. I will only add that at a period even anterior to embryonic life, namely, during the egg state, varieties appear in size and colour (as with the Hertfordshire duck with blackish eggs[2]) which reappear in the egg; in plants also the capsule and membranes of the seed are very variable and inheritable.

If then the two following propositions are admitted (and I think the first can hardly be doubted), viz. that variation of structure takes place at all times of life, though no doubt far less in amount and seldomer in quite mature life[3] (and then generally

[1] *Origin*, Ed. i. p. 444, vi. p. 610.

[2] In *Var. under Dom.*, Ed. ii. vol. I. p. 295, such eggs are said to be laid early in each season by the black Labrador duck. In the next sentence in the text the author does not distinguish the characters of the vegetable capsule from those of the ovum.

[3] This seems to me to be more strongly stated here than in the *Origin*, Ed. i.

taking the form of disease); and secondly, that these variations tend to reappear at a corresponding period of life, which seems at least probable, then we might *a priori* have expected that in any selected breed the *young* animal would not partake in a corresponding degree the peculiarities characterising the *full-grown* parent; though it would in a lesser degree. For during the thousand or ten thousand selections of slight increments in the length of the limbs of individuals necessary to produce a long-limbed breed, we might expect that such increments would take place in different individuals (as we do not certainly know at what period they do take place), some earlier and some later in the embryonic state, and some during early youth; and these increments would reappear in their offspring only at corresponding periods. Hence, the entire length of limb in the new long-limbed breed would only be acquired at the latest period of life, when that one which was latest of the thousand primary increments of length supervened. Consequently, the fœtus of the new breed during the earlier part of its existence would remain much less changed in the proportions of its limbs; and the earlier the period the less would the change be.

Whatever may be thought of the facts on which this reasoning is grounded, it shows how the embryos and young of different species might come to remain less changed than their mature parents; and practically we find that the young of our domestic animals, though differing, differ less than their full-grown parents. Thus if we look at the young puppies[1] of the greyhound and bulldog—(the two most obviously modified of the breeds of dog)—we find their puppies at the age of six days with legs and noses (the latter measured from the eyes to the tip) of the

[1] *Origin*, Ed. i. p. 444, vi. p. 611.

same length; though in the proportional thicknesses and general appearance of these parts there is a great difference. So it is with cattle, though the young calves of different breeds are easily recognisable, yet they do not differ so much in their proportions as the full-grown animals. We see this clearly in the fact that it shows the highest skill to select the best forms early in life, either in horses, cattle or poultry; no one would attempt it only a few hours after birth; and it requires great discrimination to judge with accuracy even during their full youth, and the best judges are sometimes deceived. This shows that the ultimate proportions of the body are not acquired till near mature age. If I had collected sufficient facts to firmly establish the proposition that in artificially selected breeds the embryonic and young animals are not changed in a corresponding degree with their mature parents, I might have omitted all the foregoing reasoning and the attempts to explain how this happens; for we might safely have transferred the proposition to the breeds or species naturally selected; and the ultimate effect would necessarily have been that in a number of races or species descended from a common stock and forming several genera and families the embryos would have resembled each other more closely than full-grown animals. Whatever may have been the form or habits of the parent-stock of the Vertebrata, in whatever course the arteries ran and branched, the selection of variations, supervening after the first formation of the arteries in the embryo, would not tend from variations supervening at corresponding periods to alter their course at that period: hence, the similar course of the arteries in the mammal, bird, reptile and fish, must be looked at as a most ancient record of the embryonic structure of the common parent-stock of these four great classes.

A long course of selection might cause a form to become more simple, as well as more complicated; thus the adaptation of a crustaceous[1] animal to live attached during its whole life to the body of a fish, might permit with advantage great simplification of structure, and on this view the singular fact of an embryo being more complex than its parent is at once explained.

On the graduated complexity in each great class.

I may take this opportunity of remarking that naturalists have observed that in most of the great classes a series exists from very complicated to very simple beings; thus in Fish, what a range there is between the sand-eel and shark,—in the Articulata, between the common crab and the Daphnia[2],—between the Aphis and butterfly, and between a mite and a spider[3]. Now the observation just made, namely, that selection might tend to simplify, as well as to complicate, explains this; for we can see that during the endless geologico-geographical changes, and consequent isolation of species, a station occupied in other districts by less complicated animals might be left unfilled, and be occupied by a degraded form of a higher or more complicated class; and it would by no means follow that, when the two regions became united, the degraded organism would give way to the aboriginally lower organism. According to our theory, there is obviously no power tending constantly to exalt species, except the mutual struggle between the different individuals and classes; but from the strong and general hereditary tendency we might expect to find some tendency to progressive complication in the successive production of new organic forms.

[1] *Origin*, Ed. i. p. 441, vi. p. 607.
[2] Compare *Origin*, Ed. i. p. 419, vi. p. 575.
[3] ⟨Note in original.⟩ Scarcely possible to distinguish between non-development and retrograde development.

Modification by selection of the forms of immature animals.

I have above remarked that the feline[1] form is quite of secondary importance to the embryo and to the kitten. Of course, during any great and prolonged change of structure in the mature animal, it might, and often would be, indispensable that the form of the embryo should be changed; and this could be effected, owing to the hereditary tendency at corresponding ages, by selection, equally well as in mature age: thus if the embryo tended to become, or to remain, either over its whole body or in certain parts, too bulky, the female parent would die or suffer more during parturition; and as in the case of the calves with large hinder quarters[2], the peculiarity must be either eliminated or the species become extinct. Where an embryonic form has to seek its own food, its structure and adaptation is just as important to the species as that of the full-grown animal; and as we have seen that a peculiarity appearing in a caterpillar (or in a child, as shown by the hereditariness of peculiarities in the milk-teeth) reappears in its offspring, so we can at once see that our common principle of the selection of slight accidental variations would modify and adapt a caterpillar to a new or changing condition, precisely as in the full-grown butterfly. Hence probably it is that caterpillars of different species of the Lepidoptera differ more than those embryos, at a corresponding early period of life, do which remain inactive in the womb of their parents. The parent during successive ages continuing to be adapted by selection for some one object, and the larva for quite another one, we need not wonder at

[1] See p. 42, where the same illustration is used.
[2] *Var. under Dom.*, Ed. ii. vol. I. p. 452.

the difference becoming wonderfully great between them; even as great as that between the fixed rock-barnacle and its free, crab-like offspring, which is furnished with eyes and well-articulated, locomotive limbs[1].

Importance of embryology in classification.

We are now prepared to perceive why the study of embryonic forms is of such acknowledged importance in classification[2]. For we have seen that a variation, supervening at any time, may aid in the modification and adaptation of the full-grown being; but for the modification of the embryo, only the variations which supervene at a very early period can be seized on and perpetuated by selection: hence there will be less power and less tendency (for the structure of the embryo is mostly unimportant) to modify the young: and hence we might expect to find at this period similarities preserved between different groups of species which had been obscured and quite lost in the full-grown animals. I conceive on the view of separate creations it would be impossible to offer any explanation of the affinities of organic beings thus being plainest and of the greatest importance at that period of life when their structure is not adapted to the final part they have to play in the economy of nature.

Order in time in which the great classes have first appeared.

It follows strictly from the above reasoning only that the embryos of (for instance) existing vertebrata resemble more closely the embryo of the parent-stock of this great class than do full-grown existing vertebrata resemble their full-grown parent-

[1] *Origin*, Ed. i. p. 441, vi. p. 607.
[2] *Origin*, Ed. i. p. 449, vi. p. 617.

stock. But it may be argued with much probability that in the earliest and simplest condition of things the parent and embryo must have resembled each other, and that the passage of any animal through embryonic states in its growth is entirely due to subsequent variations affecting *only* the more mature periods of life. If so, the embryos of the existing vertebrata will shadow forth the full-grown structure of some of those forms of this great class which existed at the earlier periods of the earth's history[1]: and accordingly, animals with a fish-like structure ought to have preceded birds and mammals; and of fish, that higher organized division with the vertebræ extending into one division of the tail ought to have preceded the equal-tailed, because the embryos of the latter have an unequal tail; and of Crustacea, entomostraca ought to have preceded the ordinary crabs and barnacles—polypes ought to have preceded jelly-fish, and infusorial animalcules to have existed before both. This order of precedence in time in some of these cases is believed to hold good; but I think our evidence is so exceedingly incomplete regarding the number and kinds of organisms which have existed during all, especially the earlier, periods of the earth's history, that I should put no stress on this accordance, even if it held truer than it probably does in our present state of knowledge.

[1] *Origin*, Ed. i. p. 449, vi. p. 618.

CHAPTER IX

ABORTIVE OR RUDIMENTARY ORGANS

The abortive organs of naturalists.

PARTS of structure are said to be "abortive," or when in a still lower state of development "rudimentary[1]," when the same reasoning power, which convinces us that in some cases similar parts are beautifully adapted to certain ends, declares that in others they are absolutely useless. Thus the rhinoceros, the whale[2], etc., have, when young, small but properly formed teeth, which never protrude from the jaws; certain bones, and even the entire extremities are represented by mere little cylinders or points of bone, often soldered to other bones: many beetles have exceedingly minute but regularly formed wings lying under their wing-cases[3], which latter are united never to be opened: many plants have, instead of stamens, mere filaments or little knobs; petals are reduced to scales, and whole flowers to buds, which (as in the feather hyacinth) never expand. Similar instances are almost innumerable, and are justly considered wonderful: probably not one organic being exists in which some part does not bear the stamp of inutility; for what can be clearer[4], as far as our reasoning powers

[1] In the *Origin*, Ed. i. p. 450, vi. p. 619, the author does not lay stress on any distinction in meaning between the terms *abortive* and *rudimentary* organs.

[2] *Origin*, Ed. i. p. 450, vi. p. 619.

[3] *Ibid.*

[4] This argument occurs in *Origin*, Ed. i. p. 451, vi. p. 619.

can reach, than that teeth are for eating, extremities for locomotion, wings for flight, stamens and the entire flower for reproduction; yet for these clear ends the parts in question are manifestly unfit. Abortive organs are often said to be mere representatives (a metaphorical expression) of similar parts in other organic beings; but in some cases they are more than representatives, for they seem to be the actual organ not fully grown or developed; thus the existence of mammæ in the male vertebrata is one of the oftenest adduced cases of abortion; but we know that these organs in man (and in the bull) have performed their proper function and secreted milk: the cow has normally four mammæ and two abortive ones, but these latter in some instances are largely developed and even (??) give milk[1]. Again in flowers, the representatives of stamens and pistils can be traced to be really these parts not developed; Kölreuter has shown by crossing a diæcious plant (a Cucubalus) having a rudimentary pistil[2] with another species having this organ perfect, that in the hybrid offspring the rudimentary part is more developed, though still remaining abortive; now this shows how intimately related in nature the mere rudiment and the fully developed pistil must be.

Abortive organs, which must be considered as useless as far as their ordinary and normal purpose is concerned, are sometimes adapted to other ends[3]: thus the marsupial bones, which properly serve to support the young in the mother's pouch, are present in the male and serve as the fulcrum for muscles connected only with male functions: in the

[1] *Origin*, Ed. i. p. 451, vi. p. 619, on male mammæ. In the *Origin* he speaks certainly of the abortive mammæ of the cow giving milk,—a point which is here queried.

[2] *Origin*, Ed. i. p. 451, vi. p. 620.

[3] The care of rudimentary organs adapted to new purposes is discussed in the *Origin*, Ed. i. p. 451, vi. p. 620.

male of the marigold flower the pistil is abortive for its proper end of being impregnated, but serves to sweep the pollen out of the anthers[1] ready to be borne by insects to the perfect pistils in the other florets. It is likely in many cases, yet unknown to us, that abortive organs perform some useful function; but in other cases, for instance in that of teeth embedded in the solid jaw-bone, or of mere knobs, the rudiments of stamens and pistils, the boldest imagination will hardly venture to ascribe to them any function. Abortive parts, even when wholly useless to the individual species, are of great signification in the system of nature; for they are often found to be of very high importance in a natural classification[2]; thus the presence and position of entire abortive flowers, in the grasses, cannot be overlooked in attempting to arrange them according to their true affinities. This corroborates a statement in a previous chapter, viz. that the physiological importance of a part is no index of its importance in classification. Finally, abortive organs often are only developed, proportionally with other parts, in the embryonic or young state of each species[3]; this again, especially considering the classificatory importance of abortive organs, is evidently part of the law (stated in the last chapter) that the higher affinities of organisms are often best seen in the stages towards maturity, through which the embryo passes. On the ordinary view of individual creations, I think that scarcely any class of facts in natural history are more wonderful or less capable of receiving explanation.

[1] This is here stated on the authority of Sprengel; see also *Origin*, Ed. i. p. 452, vi. p. 621.

[2] *Origin*, Ed. i. p. 455, vi. p. 627. In the margin R. Brown's name is given apparently as the authority for the fact.

[3] *Origin*, Ed. i. p. 455, vi. p. 626.

The abortive organs of physiologists.

Physiologists and medical men apply the term "abortive" in a somewhat different sense from naturalists; and their application is probably the primary one; namely, to parts, which from accident or disease before birth are not developed or do not grow[1]: thus, when a young animal is born with a little stump in the place of a finger or of the whole extremity, or with a little button instead of a head, or with a mere bead of bony matter instead of a tooth, or with a stump instead of a tail, these parts are said to be aborted. Naturalists on the other hand, as we have seen, apply this term to parts not stunted during the growth of the embryo, but which are as regularly produced in successive generations as any other most essential parts of the structure of the individual: naturalists, therefore, use this term in a metaphorical sense. These two classes of facts, however, blend into each other[2]; by parts accidentally aborted, during the embryonic life of one individual, becoming hereditary in the succeeding generations: thus a cat or dog, born with a stump instead of a tail, tends to transmit stumps to their offspring; and so it is with stumps representing the extremities; and so again with flowers, with defective and rudimentary parts, which are annually produced in new flower-buds and even in successive seedlings. The strong hereditary tendency to reproduce every either congenital or slowly acquired structure, whether useful or injurious to the individual, has been shown in the first part; so that we need feel no surprise at these truly abortive

[1] *Origin*, Ed. i. p. 454, vi. p. 625.

[2] In the *Origin*, Ed. i. p. 454, vi. p. 625, the author in referring to semi-monstrous variations adds "But I doubt whether any of these cases throw light on the origin of rudimentary organs in a state of nature." In 1844 he was clearly more inclined to an opposite opinion.

parts becoming hereditary. A curious instance of the force of hereditariness is sometimes seen in two little loose hanging horns, quite useless as far as the function of a horn is concerned, which are produced in hornless races of our domestic cattle[1]. Now I believe no real distinction can be drawn between a stump representing a tail or a horn or the extremities; or a short shrivelled stamen without any pollen; or a dimple in a petal representing a nectary, when such rudiments are regularly reproduced in a race or family, and the true abortive organs of naturalists. And if we had reason to believe (which I think we have not) that all abortive organs had been at some period *suddenly* produced during the embryonic life of an individual, and afterwards become inherited, we should at once have a simple explanation of the origin of abortive and rudimentary organs[2]. In the same manner as during changes of pronunciation certain letters in a word may become useless[3] in pronouncing it, but yet may aid us in searching for its derivation, so we can see that rudimentary organs, no longer useful to the individual, may be of high importance in ascertaining its descent, that is, its true classification in the natural system.

Abortion from gradual disuse.

There seems to be some probability that continued disuse of any part or organ, and the selection of individuals with such parts slightly less developed, would in the course of ages produce in

[1] *Origin*, Ed. i. p. 454, vi. p. 625.

[2] See *Origin*, Ed. i. p. 454, vi. p. 625. The author there discusses monstrosities in relation to rudimentary organs, and comes to the conclusion that disuse is of more importance, giving as a reason his doubt "whether species under nature ever undergo abrupt changes." It seems to me that in the *Origin* he gives more weight to the "Lamarckian factor" than he did in 1844. Huxley took the opposite view, see the Introduction.

[3] *Origin*, Ed. i. p. 455, vi. p. 627.

organic beings under domesticity races with such parts abortive. We have every reason to believe that every part and organ in an individual becomes fully developed only with exercise of its functions; that it becomes developed in a somewhat lesser degree with less exercise; and if forcibly precluded from all action, such part will often become atrophied. Every peculiarity, let it be remembered, tends, especially where both parents have it, to be inherited. The less power of flight in the common duck compared with the wild, must be partly attributed to disuse[1] during successive generations, and as the wing is properly adapted to flight, we must consider our domestic duck in the first stage towards the state of the Apteryx, in which the wings are so curiously abortive. Some naturalists have attributed (and possibly with truth) the falling ears so characteristic of most domestic dogs, some rabbits, oxen, cats, goats, horses, &c., &c., as the effects of the lesser use of the muscles of these flexible parts during successive generations of inactive life; and muscles, which cannot perform their functions, must be considered verging towards abortion. In flowers, again, we see the gradual abortion during successive seedlings (though this is more properly a conversion) of stamens into imperfect petals, and finally into perfect petals. When the eye is blinded in early life the optic nerve sometimes becomes atrophied; may we not believe that where this organ, as is the case with the subterranean mole-like Tuco-tuco ⟨*Ctenomys*⟩[2], is frequently impaired and lost, that in the course of generations the whole organ might become abortive, as it normally is in some burrowing quadrupeds having nearly similar habits with the Tuco-tuco?

[1] *Origin*, Ed. i. p. 11, vi. p. 13, where drooping-ears of domestic animals are also given.

[2] *Origin*, Ed. i. p. 137, vi. p. 170.

In as far then as it is admitted as probable that the effects of disuse (together with occasional true and sudden abortions during the embryonic period) would cause a part to be less developed, and finally to become abortive and useless; then during the infinitely numerous changes of habits in the many descendants from a common stock, we might fairly have expected that cases of organs becom⟨ing⟩ abortive would have been numerous. The preservation of the stump of the tail, as usually happens when an animal is born tailless, we can only explain by the strength of the hereditary principle and by the period in embryo when affected[1]: but on the theory of disuse gradually obliterating a part, we can see, according to the principles explained in the last chapter (viz. of hereditariness at corresponding periods of life[2], together with the use and disuse of the part in question not being brought into play in early or embryonic life), that organs or parts would tend not to be utterly obliterated, but to be reduced to that state in which they existed in early embryonic life. Owen often speaks of a part in a full-grown animal being in an "embryonic condition." Moreover we can thus see why abortive organs are most developed at an early period of life. Again, by gradual selection, we can see how an organ rendered abortive in its primary use might be converted to other purposes; a duck's wing might come to serve for a fin, as does that of the penguin; an abortive bone might come to serve, by the slow increment and change of place in the muscular fibres, as a fulcrum for a new series of muscles; the pistil[3] of the marigold might become abortive as a reproductive part, but be continued in its function of sweeping the pollen out of the anthers; for if in

[1] These words seem to have been inserted as an afterthought.
[2] *Origin*, Ed. i. p. 444, vi. p. 611.
[3] This and similar cases occur in the *Origin*, Ed. i. p. 452, vi. p. 621.

this latter respect the abortion had not been checked by selection, the species must have become extinct from the pollen remaining enclosed in the capsules of the anthers.

Finally then I must repeat that these wonderful facts of organs formed with traces of exquisite care, but now either absolutely useless or adapted to ends wholly different from their ordinary end, being present and forming part of the structure of almost every inhabitant of this world, both in long-past and present times—being best developed and often only discoverable at a very early embryonic period, and being full of signification in arranging the long series of organic beings in a natural system—these wonderful facts not only receive a simple explanation on the theory of long-continued selection of many species from a few common parent-stocks, but necessarily follow from this theory. If this theory be rejected, these facts remain quite inexplicable; without indeed we rank as an explanation such loose metaphors as that of De Candolle's[1], in which the kingdom of nature is compared to a well-covered table, and the abortive organs are considered as put in for the sake of symmetry!

[1] The metaphor of the dishes is given in the Essay of 1842, p. 47, note 3.

CHAPTER X

RECAPITULATION AND CONCLUSION

Recapitulation.

I WILL now recapitulate the course of this work, more fully with respect to the former parts, and briefly ⟨as to⟩ the latter. In the first chapter we have seen that most, if not all, organic beings, when taken by man out of their natural condition, and bred during several generations, vary; that is variation is partly due to the direct effect of the new external influences, and partly to the indirect effect on the reproductive system rendering the organization of the offspring in some degree plastic. Of the variations thus produced, man when uncivilised naturally preserves the life, and therefore unintentionally breeds from those individuals most useful to him in his different states: when even semi-civilised, he intentionally separates and breeds from such individuals. Every part of the structure seems occasionally to vary in a very slight degree, and the extent to which all kinds of peculiarities in mind and body, when congenital and when slowly acquired either from external influences, from exercise, or from disuse ⟨are inherited⟩, is truly wonderful. When several breeds are once formed, then crossing is the most fertile source of new breeds[1]. Variation

[1] Compare however Darwin's later view:—"The possibility of making distinct races by crossing has been greatly exaggerated," *Origin*, Ed. i. p. 20, vi. p. 23. The author's change of opinion was no doubt partly due to his experience in breeding pigeons.

must be ruled, of course, by the health of the new race, by the tendency to return to the ancestral forms, and by unknown laws determining the proportional increase and symmetry of the body. The amount of variation, which has been effected under domestication, is quite unknown in the majority of domestic beings.

In the second chapter it was shown that wild organisms undoubtedly vary in some slight degree: and that the kind of variation, though much less in degree, is similar to that of domestic organisms. It is highly probable that every organic being, if subjected during several generations to new and varying conditions, would vary. It is certain that organisms, living in an *isolated* country which is undergoing geological changes, must in the course of time be so subjected to new conditions; moreover an organism, when by chance transported into a new station, for instance into an island, will often be exposed to new conditions, and be surrounded by a new series of organic beings. If there were no power at work selecting every slight variation, which opened new sources of subsistence to a being thus situated, the effects of crossing, the chance of death and the constant tendency to reversion to the old parent-form, would prevent the production of new races. If there were any selective agency at work, it seems impossible to assign any limit[1] to the complexity and beauty of the adaptive structures, which *might* thus be produced: for certainly the limit of possible variation of organic beings, either in a wild or domestic state, is not known.

It was then shown, from the geometrically increasing tendency of each species to multiply (as evidenced from what we know of mankind and

[1] In the *Origin*, Ed. i. p. 469, vi. p. 644, Darwin makes a strong statement to this effect.

of other animals when favoured by circumstances), and from the means of subsistence of each species on an *average* remaining constant, that during some part of the life of each, or during every few generations, there must be a severe struggle for existence; and that less than a grain[1] in the balance will determine which individuals shall live and which perish. In a country, therefore, undergoing changes, and cut off from the free immigration of species better adapted to the new station and conditions, it cannot be doubted that there is a most powerful means of selection, *tending* to preserve even the slightest variation, which aided the subsistence or defence of those organic beings, during any part of their whole existence, whose organization had been rendered plastic. Moreover, in animals in which the sexes are distinct, there is a sexual struggle, by which the most vigorous, and consequently the best adapted, will oftener procreate their kind.

A new race thus formed by natural selection would be undistinguishable from a species. For comparing, on the one hand, the several species of a genus, and on the other hand several domestic races from a common stock, we cannot discriminate them by the amount of external difference, but only, first, by domestic races not remaining so constant or being so "true" as species are; and secondly by races always producing fertile offspring when crossed. And it was then shown that a race naturally selected—from the variation being slower —from the selection steadily leading towards the same ends[2], and from every new slight change in structure being adapted (as is implied by its selec-

[1] "A grain in the balance will determine which individual shall live and which shall die," *Origin*, Ed. i. p. 467, vi. p. 642. A similar statement occurs in the 1842 Essay, p. 8, note 3.

[2] Thus according to the author what is now known as *orthogenesis* is due to selection.

tion) to the new conditions and being fully exercised, and lastly from the freedom from occasional crosses with other species, would almost necessarily be "truer" than a race selected by ignorant or capricious and short-lived man. With respect to the sterility of species when crossed, it was shown not to be a universal character, and when present to vary in degree: sterility also was shown probably to depend less on external than on constitutional differences. And it was shown that when individual animals and plants are placed under new conditions, they become, without losing their healths, as sterile, in the same manner and to the same degree, as hybrids; and it is therefore conceivable that the cross-bred offspring between two species, having different constitutions, might have its constitution affected in the same peculiar manner as when an individual animal or plant is placed under new conditions. Man in selecting domestic races has little wish and still less power to adapt the whole frame to new conditions; in nature, however, where each species survives by a struggle against other species and external nature, the result must be very different.

Races descending from the same stock were then compared with species of the same genus, and they were found to present some striking analogies. The offspring also of races when crossed, that is mongrels, were compared with the cross-bred offspring of species, that is hybrids, and they were found to resemble each other in all their characters, with the one exception of sterility, and even this, when present, often becomes after some generations variable in degree. The chapter was summed up, and it was shown that no ascertained limit to the amount of variation is known; or could be predicted with due time and changes of condition granted. It was then admitted that although the production of new races, undistinguishable from true species, is probable, we must look to the relations in the past and present geographical distribution of the infinitely numerous beings, by which we are surrounded—to their affinities and to their structure—for any direct evidence.

In the third chapter the inheritable variations in the mental phenomena of domestic and of wild organic beings were considered. It was shown that we are not concerned in this work with the first origin of the leading mental qualities; but that tastes, passions, dispositions, consensual movements, and habits all became, either congenitally or during mature life, modified and were inherited. Several of these modified habits were found to correspond in every essential character with true instincts, and they were found to follow the same laws. Instincts and dispositions &c. are fully as important to the preservation and increase of a species as its corporeal structure; and therefore the natural means of selection would act on and modify them equally with corporeal structures. This being granted, as well as the proposition that mental phenomena are variable, and that the modifications are inheritable, the possibility of the several most complicated instincts being slowly acquired was considered, and it was shown from the very imperfect series in the instincts of the animals now existing, that we are not justified in *prima facie* rejecting a theory of the common descent of allied organisms from the difficulty of imagining the transitional stages in the various now most complicated and wonderful instincts. We were thus led on to consider the same question with respect both to highly complicated organs, and to the aggregate of several such organs, that is individual organic beings; and it was shown, by the same method of taking the existing most imperfect series, that we ought not at once to reject the theory, because we cannot trace the transitional stages in such organs, or conjecture the transitional habits of such individual species.

In the Second Part[1] the direct evidence of allied forms having descended from the same stock was discussed. It was shown that this theory requires a long series of intermediate forms between the species and groups in the same classes—forms not directly intermediate between existing species, but intermediate with a common parent. It was admitted that if even all the preserved fossils and existing species were collected, such a series would be far from being formed; but it was shown that we have not *good* evidence that the oldest known deposits are contemporaneous with the first appearance of living beings; or that the several subsequent formations are nearly consecutive; or that any one formation preserves a nearly perfect fauna of even the hard marine organisms, which lived in that quarter of the world. Consequently, we have no reason to suppose that more than a small fraction of the organisms which have lived at any one period have ever been preserved; and hence that we ought not to expect to discover the fossilised sub-varieties between any two species. On the other hand, the evidence, though extremely imperfect, drawn from fossil remains, as far as it does go, is in favour of such a series of organisms having existed as that required. This want of evidence of the past existence of almost infinitely numerous intermediate forms, is, I conceive, much the weightiest difficulty[2] on the theory of common descent; but I must think that this is due to ignorance necessarily resulting from the imperfection of all geological records.

[1] Part II begins with Ch. IV. See the Introduction, where the absence of division into two parts (in the *Origin*) is discussed.

[2] In the recapitulation in the last chapter of the *Origin*, Ed. i. p. 475, vi. p. 651, the author does not insist on this point as the weightiest difficulty, though he does so in Ed. i. p. 299. It is possible that he had come to think less of the difficulty in question: this was certainly the case when he wrote the 6th edition, see p. 438.

In the fifth chapter it was shown that new species gradually[1] appear, and that the old ones gradually disappear, from the earth; and this strictly accords with our theory. The extinction of species seems to be preceded by their rarity; and if this be so, no one ought to feel more surprise at a species being exterminated than at its being rare. Every species which is not increasing in number must have its geometrical tendency to increase checked by some agency seldom accurately perceived by us. Each slight increase in the power of this unseen checking agency would cause a corresponding decrease in the average numbers of that species, and the species would become rarer: we feel not the least surprise at one species of a genus being rare and another abundant; why then should we be surprised at its extinction, when we have good reason to believe that this very rarity is its regular precursor and cause.

In the sixth chapter the leading facts in the geographical distribution of organic beings were considered—namely, the dissimilarity in areas widely and effectually separated, of the organic beings being exposed to very similar conditions (as for instance, within the tropical forests of Africa and America, or on the volcanic islands adjoining them). Also the striking similarity and general relations of the inhabitants of the same great continents, conjoined with a lesser degree of dissimilarity in the inhabitants living on opposite sides of the barriers intersecting it—whether or not these opposite sides are exposed to similar conditions. Also the dissimilarity, though in a still lesser degree, in the inhabitants of different islands in the same archipelago, together with their similarity taken as a

[1] ⟨The following words :⟩ The fauna changes singly ⟨were inserted by the author, apparently to replace a doubtful erasure⟩.

whole with the inhabitants of the nearest continent, whatever its character may be. Again, the peculiar relations of Alpine floras; the absence of mammifers on the smaller isolated islands; and the comparative fewness of the plants and other organisms on islands with diversified stations; the connection between the possibility of occasional transportal from one country to another, with an affinity, though not identity, of the organic beings inhabiting them. And lastly, the clear and striking relations between the living and the extinct in the same great divisions of the world; which relation, if we look very far backward, seems to die away. These facts, if we bear in mind the geological changes in progress, all simply follow from the proposition of allied organic beings having lineally descended from common parent-stocks. On the theory of independent creations they must remain, though evidently connected together, inexplicable and disconnected.

In the seventh chapter, the relationship or grouping of extinct and recent species; the appearance and disappearance of groups; the ill-defined objects of the natural classification, not depending on the similarity of organs physiologically important, not being influenced by adaptive or analogical characters, though these often govern the whole economy of the individual, but depending on any character which varies least, and especially on the forms through which the embryo passes, and, as was afterwards shown, on the presence of rudimentary and useless organs. The alliance between the nearest species in *distinct* groups being general and not especial; the close similarity in the rules and objects in classifying domestic races and true species. All these facts were shown to follow on the natural system being a genealogical system.

In the eighth chapter, the unity of structure throughout large groups, in species adapted to the most different lives, and the wonderful metamorphosis (used metaphorically by naturalists) of one part or organ into another, were shown to follow simply on new species being produced by the selection and inheritance of successive *small* changes of structure. The unity of type is wonderfully manifested by the similarity of structure, during the embryonic period, in the species of entire classes. To explain this it was shown that the different races of our domestic animals differ less, during their young state, than when full grown; and consequently, if species are produced like races, the same fact, on a greater scale, might have been expected to hold good with them. This remarkable law of nature was attempted to be explained through establishing, by sundry facts, that slight variations originally appear during all periods of life, and that when inherited they tend to appear at the corresponding period of life; according to these principles, in several species descended from the same parent-stock, their embryos would almost necessarily much more closely resemble each other than they would in their adult state. The importance of these embryonic resemblances, in making out a natural or genealogical classification, thus becomes at once obvious. The occasional greater simplicity of structure in the mature animal than in the embryo; the gradation in complexity of the species in the great classes; the adaptation of the larvæ of animals to independent powers of existence; the immense difference in certain animals in their larval and mature states, were all shown on the above principles to present no difficulty.

In the ⟨ninth⟩ chapter, the frequent and almost general presence of organs and parts, called by naturalists abortive or rudimentary, which, though formed with exquisite care, are generally absolutely useless ⟨was considered⟩. ⟨These structures,⟩ though sometimes applied to uses not normal,—which cannot be considered as mere representative parts, for they are sometimes capable of performing their proper function,—which are always best developed, and sometimes only developed, during a very early period of life,—and which are of admitted high importance in classification,—were shown to be simply explicable on our theory of common descent.

Why do we wish to reject the theory of common descent?

Thus have many general facts, or laws, been included under one explanation; and the difficulties encountered are those which would naturally result from our acknowledged ignorance. And why should we not admit this theory of descent[1]? Can it be shown that organic beings in a natural state are *all absolutely invariable?* Can it be said that the *limit of variation* or the number of varieties capable of being formed under domestication are known? Can any distinct line be drawn *between a race and a species?* To these three questions we may certainly answer in the negative. As long as species were thought to be divided and defined by an impassable barrier of *sterility*, whilst we were ignorant of geology, and imagined that the *world was of short duration*, and the number of its past inhabitants few, we were justified in assuming individual creations, or in saying with Whewell that the beginnings of all things are hidden from man. Why then do we feel so strong an inclination to reject this theory —especially when the actual case of any two species, or even of any two races, is adduced—and one is asked, have these two originally descended from the same parent womb? I believe it is because we are

[1] This question forms the subject of what is practically a section of the final chapter of the *Origin* (Ed. i. p. 480, vi. p. 657).

always slow in admitting any great change of which we do not see the intermediate steps. The mind cannot grasp the full meaning of the term of a million or hundred million years, and cannot consequently add up and perceive the full effects of small successive variations accumulated during almost infinitely many generations. The difficulty is the same with that which, with most geologists, it has taken long years to remove, as when Lyell propounded that great valleys[1] were hollowed out [and long lines of inland cliffs had been formed] by the slow action of the waves of the sea. A man may long view a grand precipice without actually believing, though he may not deny it, that thousands of feet in thickness of solid rock once extended over many square miles where the open sea now rolls; without fully believing that the same sea which he sees beating the rock at his feet has been the sole removing power.

Shall we then allow that the three distinct species of rhinoceros[2] which separately inhabit Java and Sumatra and the neighbouring mainland of Malacca were created, male and female, out of the inorganic materials of these countries? Without any adequate cause, as far as our reason serves, shall we say that they were merely, from living near each other, created very like each other, so as to form a section of the genus dissimilar from the African section, some of the species of which section inhabit very similar and some very dissimilar stations? Shall we say that without any apparent cause they were created on the same generic type with the ancient woolly rhinoceros of Siberia and of the other species which formerly inhabited the same main division of the world: that they were created, less

[1] *Origin*, Ed. i. p. 481, vi. p. 659.

[2] The discussion on the three species of *Rhinoceros* which also occurs in the Essay of 1842, p. 48, was omitted in Ch. XIV of the *Origin*, Ed. i.

and less closely related, but still with interbranching affinities, with all the other living and extinct mammalia? That without any apparant adequate cause their short necks should contain the same number of vertebræ with the giraffe; that their thick legs should be built on the same plan with those of the antelope, of the mouse, of the hand of the monkey, of the wing of the bat, and of the fin of the porpoise. That in each of these species the second bone of their leg should show clear traces of two bones having been soldered and united into one; that the complicated bones of their head should become intelligible on the supposition of their having been formed of three expanded vertebræ; that in the jaws of each when dissected young there should exist small teeth which never come to the surface. That in possessing these useless abortive teeth, and in other characters, these three rhinoceroses in their embryonic state should much more closely resemble other mammalia than they do when mature. And lastly, that in a still earlier period of life, their arteries should run and branch as in a fish, to carry the blood to gills which do not exist. Now these three species of rhinoceros closely resemble each other; more closely than many generally acknowledged races of our domestic animals; these three species if domesticated would almost certainly vary, and races adapted to different ends might be selected out of such variations. In this state they would probably breed together, and their offspring would possibly be quite, and probably in some degree, fertile; and in either case, by continued crossing, one of these specific forms might be absorbed and lost in another. I repeat, shall we then say that a pair, or a gravid female, of each of these three species of rhinoceros, were separately created with deceptive appearances of true relationship, with the stamp of inutility on some parts, and of conversion in other parts, out of the inorganic elements of Java, Sumatra and Malacca? or have they descended, like our domestic races, from the same parent-stock? For my own part I could no more admit the former proposition than I could admit that the planets move in their courses, and that a stone falls to the ground, not through the intervention of the secondary and appointed law of gravity, but from the direct volition of the Creator.

Before concluding it will be well to show, although this has incidentally appeared, how far the theory of common descent can legitimately be extended[1]. If we once admit that two true species of the same genus can have descended from the same parent, it will not be possible to deny that two species of two genera may also have descended from a common stock. For in some families the genera approach almost as closely as species of the same genus; and in some orders, for instance in the monocotyledonous plants, the families run closely into each other. We do not hesitate to assign a common origin to dogs or cabbages, because they are divided into groups analogous to the groups in nature. Many naturalists indeed admit that all groups are artificial; and that they depend entirely on the extinction of intermediate species. Some naturalists, however, affirm that though driven from considering sterility as the characteristic of species, that an entire incapacity to propagate together is the best evidence of the existence of natural genera. Even if we put on one side the undoubted fact that some species of the same genus

[1] This corresponds to a paragraph in the *Origin*, Ed. i. p. 483, vi. p. 662, where it is assumed that animals have descended "from at most only four or five progenitors, and plants from an equal or lesser number." In the *Origin*, however, the author goes on, Ed. i. p. 484, vi. p. 663: "Analogy would lead me one step further, namely, to the belief that all animals and plants have descended from some one prototype."

will not breed together, we cannot possibly admit the above rule, seeing that the grouse and pheasant (considered by some good ornithologists as forming two families), the bull-finch and canary-bird have bred together.

No doubt the more remote two species are from each other, the weaker the arguments become in favour of their common descent. In species of two distinct families the analogy, from the variation of domestic organisms and from the manner of their intermarrying, fails; and the arguments from their geographical distribution quite or almost quite fails. But if we once admit the general principles of this work, as far as a clear unity of type can be made out in groups of species, adapted to play diversified parts in the economy of nature, whether shown in the structure of the embryonic or mature being, and especially if shown by a community of abortive parts, we are legitimately led to admit their community of descent. Naturalists dispute how widely this unity of type extends: most, however, admit that the vertebrata are built on one type; the articulata on another; the mollusca on a third; and the radiata on probably more than one. Plants also appear to fall under three or four great types. On this theory, therefore, all the organisms *yet discovered* are descendants of probably less than ten parent-forms.

Conclusion.

My reasons have now been assigned for believing that specific forms are not immutable creations[1]. The terms used by naturalists of affinity, unity of type, adaptive characters, the metamorphosis and

[1] This sentence corresponds, not to the final section of the *Origin*, Ed. i. p. 484, vi. p. 664, but rather to the opening words of the section already referred to (*Origin*, Ed. i. p. 480, vi. p. 657).

abortion of organs, cease to be metaphorical expressions and become intelligible facts. We no longer look at an organic being as a savage does at a ship[1] or other great work of art, as at a thing wholly beyond his comprehension, but as a production that has a history which we may search into. How interesting do all instincts become when we speculate on their origin as hereditary habits, or as slight congenital modifications of former instincts perpetuated by the individuals so characterised having been preserved. When we look at every complex instinct and mechanism as the summing up of a long history of contrivances, each most useful to its possessor, nearly in the same way as when we look at a great mechanical invention as the summing up of the labour, the experience, the reason, and even the blunders of numerous workmen. How interesting does the geographical distribution of all organic beings, past and present, become as throwing light on the ancient geography of the world. Geology loses glory[2] from the imperfection of its archives, but it gains in the immensity of its subject. There is much grandeur in looking at every existing organic being either as the lineal successor of some form now buried under thousands of feet of solid rock, or as being the co-descendant of that buried form of some more ancient and utterly lost inhabitant of this world. It accords with what we know of the laws impressed by the Creator[3] on matter that the production and extinction of forms should, like the birth and death of individuals, be

[1] This simile occurs in the Essay of 1842, p. 50, and in the *Origin*, Ed. i. p. 485, vi. p. 665, *i.e.* in the final section of Ch. XIV (vi. Ch. XV). In the MS. there is some erasure in pencil of which I have taken no notice.

[2] An almost identical sentence occurs in the *Origin*, Ed. i. p. 487, vi. p. 667. The fine prophecy (in the *Origin*, Ed. i. p. 486, vi. p. 666) on "the almost untrodden field of inquiry" is wanting in the present Essay.

[3] See the last paragraph on p. 488 of the *Origin*, Ed. i., vi. p. 668.

the result of secondary means. It is derogatory that the Creator of countless Universes should have made by individual acts of His will the myriads of creeping parasites and worms, which since the earliest dawn of life have swarmed over the land and in the depths of the ocean. We cease to be astonished[1] that a group of animals should have been formed to lay their eggs in the bowels and flesh of other sensitive beings; that some animals should live by and even delight in cruelty; that animals should be led away by false instincts; that annually there should be an incalculable waste of the pollen, eggs and immature beings; for we see in all this the inevitable consequences of one great law, of the multiplication of organic beings not created immutable. From death, famine, and the struggle for existence, we see that the most exalted end which we are capable of conceiving, namely, the creation of the higher animals[2], has directly proceeded. Doubtless, our first impression is to disbelieve that any secondary law could produce infinitely numerous organic beings, each characterised by the most exquisite workmanship and widely extended adaptations: it at first accords better with our faculties to suppose that each required the fiat of a Creator. There[3] is a [simple] grandeur in this view of life with its several powers of growth, reproduction and of sensation, having been originally breathed into matter under a few forms, perhaps into only one[4],

[1] A passage corresponding to this occurs in the sketch of 1842, p. 51, but not in the last chapter of the *Origin*.

[2] This sentence occurs in an almost identical form in the *Origin*, Ed. i. p. 490, vi. p. 669. It will be noted that man is not named though clearly referred to. Elsewhere (*Origin*, Ed. i. p. 488) the author is bolder and writes "Light will be thrown on the origin of man and his history." In Ed. vi. p. 668, he writes "Much light &c."

[3] For the history of this sentence (with which the *Origin of Species* closes) see the Essay of 1842, p. 52, note 2: also the concluding pages of the Introduction.

[4] These four words are added in pencil between the lines.

and that whilst this planet has gone cycling onwards according to the fixed laws of gravity and whilst land and water have gone on replacing each other—that from so simple an origin, through the selection of infinitesimal varieties, endless forms most beautiful and most wonderful have been evolved.

INDEX

For the names of Authors, Birds, Mammals (including names of classes) and Plants, see sub-indexes under *Authors*, *Birds*, *Mammals* and *Plants*.

CAMBRIDGE: PRINTED BY JOHN CLAY, M.A. AT THE UNIVERSITY PRESS.

Printed in the United States
By Bookmasters